Advancing Silicon Carbide Electronics Technology I
Metal Contacts to Silicon Carbide: Physics, Technology, Applications

Edited by

Konstantinos Zekentes[1] and Konstantin Vasilevskiy[2]

[1]Foundation for Research and Technology, Heraklion, Greece

[2]School of Engineering, Newcastle University, Newcastle upon Tyne, NE1 7RU, United Kingdom

Cover illustration:

Spatial distribution of sub-micron Ni_2Si inclusions in a NiSi Schottky contact formed on 4H-SiC. The images were obtained by two-dimensional micro-Raman mapping Ni_2Si band (140 cm^{-1}) intensity normalized to the signal background. See Chapter 3 in this volume for the in-depth review of inhomogeneous SiC Schottky contacts characterisation.

The figure is taken (with minor editing) from I. Nikitina, K. Vassilevski, A. Horsfall, N. Wright, A.G. O'Neill, S.K. Ray, K. Zekentes, and C.M. Johnson, "Phase composition and electrical characteristics of nickel silicide Schottky contacts formed on 4H-SiC," *Semiconductor Science and Technology*, vol. 24, no. 5, pp. 055006, 2009. doi:10.1088/0268-1242/24/5/055006 © IOP Publishing. Reproduced with permission. All rights reserved.

Published by **Materials Research Forum LLC**
Millersville, PA 17551, USA

Published as part of the book series
Materials Research Foundations
Volume 37 (2018)
ISSN 2471-8890 (Print)
ISSN 2471-8904 (Online)

Print ISBN 978-1-945291-84-5
ePDF ISBN 978-1-945291-85-2

This book contains information obtained from authentic and highly regarded sources. Reasonable efforts have been made to publish reliable data and information, but the author and publisher cannot assume responsibility for the validity of all materials or the consequences of their use. The authors and publishers have attempted to trace the copyright holders of all material reproduced in this publication and apologize to copyright holders if permission to publish in this form has not been obtained. If any copyright material has not been acknowledged please write and let us know so we may rectify this in any future reprints.

Distributed worldwide by

Materials Research Forum LLC
105 Springdale Lane
Millersville, PA 17551
USA
http://www.mrforum.com

Manufactured in the United States of America
10 9 8 7 6 5 4 3 2 1

Table of Contents

Preface

Silicon carbide is a wide band gap semiconductor material possessing unique physical, chemical and electrical properties. Its high critical electric field and thermal conductivity make silicon carbide an excellent material for fabrication of high-power, low-loss semiconductor devices. High thermal stability, outstanding chemical inertness and hardness potentially allow SiC devices to operate in harsh environment at high-temperature conditions. Silicon carbide electronics has a rich history starting from carborundum detectors, the first commercial semiconductor devices released in 1906 [1] but shortly replaced by vacuum tubes. Development of a new growth method to produce high quality SiC crystals in the middle 50th [2] resulted in the resurrection of silicon carbide as a material for the semiconductor electronics. The start of this period was marked by the first silicon carbide conference in Boston in 1959 [3]. In the following two decades, extensive study of SiC material properties was performed and SiC processing technology was significantly improved but availability only of irregular shaped SiC platelets allowed just small scale production of blue and yellow SiC LEDs for specific applications. Seeded growth of SiC crystals was invented in the late 70th [4] and paved the way to large scale manufacturing of SiC devices. Development of SiC epitaxy in the late 80th [5] and introduction of standard size SiC wafers and blue LEDs to the market in the early 90th resulted in the renaissance of the interest to SiC as a material for commercial electronics.

Although SiC LEDs were replaced in the market by more efficient III-N LEDs in the end 90th, developed SiC growth and processing technology was not vanished. Rapid growth of using renewable energy and introduction of strict measures to reduce the carbon footprint in the last decade resulted in a great demand for highly efficient power electronics. Silicon carbide with its outstanding properties and developed growth and processing technology became a material of choice for a new generation of semiconductor power devices. Today, SiC Schottky diodes and MOSFETs are commercially available on the market and silicon carbide electronics is recognized as an essential part of modern industry occupying its own niche of highly efficient power semiconductor devices. This results in a remarkable intensification of research in SiC technology stimulated by growing demand for improving performance and reliability of SiC devices and cost efficiency of their commercial production. An additional driving force of further development of SiC technology is a great potential of silicon carbide as a material for high temperature and high frequency electronics, which is still not realized and awaiting for convincing demonstration of SiC advantageous over conventional semiconductors for these applications.

Since 1993, the progress and latest trends in the field are regular discussed at International and European Conferences on Silicon Carbide and Related Materials. Now, proceedings of these conferences are published every year and widely disseminated. At the same time, detailed reviews and in-depth analysis of current state in SiC material characterization, crystal growth and device processing technology are given in the series of hardcover books. The first book of this kind was published in 1997 [6]. It was edited by Wolfgang J. Choyke, Hiroyuki Matsunami and Gerhard Pensl who collected about 50 papers covering the whole field from SiC fundamental properties, crystal growth and material characterization through to device processing and to the design and applications of SiC devices. It was very popular in scientific community for many years and even nicknamed as a "blue book". In 2004, it received an update with the most recent developments in SiC electronics [7] and remains very relevant as an essential reference book up to now. Few more edited books were published [8-15] from that time till 2015. They provided collections of review papers in various subjects of SiC electronics selected according to editors preferences. In 2014, a classical textbook was published by Tsunenobu Kimoto and James A. Cooper [16]. In contrast to other books, it does not aim to give a complete in-depth overview of each topic but seamlessly describes silicon carbide technology from material properties to system applications of SiC devices. These are all the books published on SiC in 21st century to the best of our knowledge and we believe that a new book is right on-time to support the periodicity of publications and to review the state-of-the art in rapidly advancing SiC technology.

SiC science and technology became matured during two last decades and include many aspects from basic physics to electrical circuit design. Obviously, all of them can be barely covered in a single volume. Bearing this in mind, we limited the scope of our book mainly by SiC device processing technology. The first volume of this book is devoted to an important part of SiC device processing which is fabrication and characterization of metal contacts. The volume is opened with Chapter 1 focusing on silicon carbide surface cleaning which is the first and essential step in any device processing. It is followed by Chapter 2 describing fundamental physics, electrical characterization methods and processing of ohmic contacts to silicon carbide. The chapter provides detailed analysis of contact resistivity dependence on material properties, limitations and accuracy of contact resistivity measurements, practical advises on ohmic contact fabrication and test structure design, critical overview of different metallization schemes and processing technologies reported up to now. Thermal stability of ohmic contacts to SiC, their protection and compatibility with device processing are also discussed in this chapter. In Chapter 3, the basic physical principles of Schottky barrier formation are recalled and adapted to the specific case of SiC. Next, the important fundamental topic of Schottky barrier inhomogeneity in SiC materials is introduced. Then, a section of this chapter is devoted to the technology and design of 4H-SiC Schottky and Junction Barrier Schottky diodes. Si/SiC heterojunction diodes are also briefly discussed in this chapter as a particular case of rectifying contacts. Some common applications of SiC Schottky diodes in power electronics and temperature/light sensors are provided in the last section of this chapter. The volume is concluded with Chapter 4 reviewing high power SiC unipolar and bipolar switching devices. The challenges and prospects of different types of SiC devices including material and technology constraints on the device performance are discussed in this chapter to elucidate the main application area of metal contacts to silicon carbide.

The book should be of high interest for technologists, scientists, engineers and graduate students who are working in the field of silicon carbide and related materials. The book also can be used as a supplementary textbook for graduate courses on related specialization.

Konstantin Vasilevskiy
Konstantinos Zekentes

References

[1] H. H. C. Dunwoody, *Wireless telegraph system*, US Patent 837616, 1906.

[2] J. A. Lely, "Darstellung von Einkristallen von Silicium Carbid und Beherrschung von Art und Menge der eingebauten Verunreinigungen," *Berichte der Deutschen Keramischen Gesellschaft,* vol. 8, pp. 229, 1955.

[3] J. R. O'Connor, and C. E. Smiltens eds., "Silicon Carbide, A High Temperature Semiconductor," New York: Pergamon, 1960, p. 521.

[4] Y. M. Tairov, and V. F. Tsvetkov, "Investigation of growth processes of ingots of silicon carbide single crystals," *Journal of Crystal Growth,* vol. 43, no. 2, pp. 209-212, 1978/03, 1978.

[5] N. Kuroda, K. Shibahara, W. Yoo, S. Nishino, and H. Matsunami, "Step-Controlled VPE Growth of SiC Single Crystals at Low Temperatures," in Extended Abstracts of the 1987 Conference on Solid State Devices and Materials, 1987, pp. 227-230. https://doi.org/10.7567/SSDM.1987.C-4-2

[6]   W. J. Choyke, H. Matsunami, and G. Pensl eds., "Silicon Carbide, A Review of Fundamental Questions and of Applications to Current Device Technology," Wiley, 1997.

[7]   W. J. Choyke, H. Matsunami, and G. Pensl eds., "Silicon Carbide, Recent Major Advances," *Advanced Texts in Physics*: Springer Berlin Heidelberg, 2004, p. 899. https://doi.org/10.1007/978-3-642-18870-1

[8]   C. F. Zhe, and J. H. Zhao eds., "Silicon Carbide," CRC Press, 2003, p. 416.

[9]   S. E. Saddow, and A. K. Agarwal eds., "Advances in Silicon Carbide Processing and Applications," Artech House, 2004, p. 228.

[10]   M. Shur, S. Rumyantsev, and M. Levinshtein eds., "SiC Materials and Devices," *Selected Topics in Electronics and Systems*, 2006. https://doi.org/10.1142/6134

[11]   P. Friedrichs, T. Kimoto, L. Ley, and G. Pensl eds., "Silicon Carbide: Volume 1: Growth, Defects, and Novel Applications," Wiley, 2011, p. 528.

[12]   P. Friedrichs, T. Kimoto, L. Ley, and G. Pensl eds., "Silicon Carbide, Volume 2: Power Devices and Sensors," Wiley, 2011, p. 520.

[13]   R. Gerhardt ed. "Properties and Applications of Silicon Carbide," InTech, 2011. https://doi.org/10.5772/615

[14]   M. Mukherjee ed. "Silicon Carbide - Materials, Processing and Applications in Electronic Devices," InTech, 2011, p. 558. https://doi.org/10.5772/852

[15]   S. Saddow, and F. L. Via eds., "Advanced Silicon Carbide Devices and Processing," InTech, 2015. https://doi.org/10.5772/59734

[16]   T. Kimoto, and J. A. Cooper, *Fundamentals of Silicon Carbide Technology*: John Wiley & Sons Singapore Pte. Ltd, 2014.

Advancing Silicon Carbide Electronics Technology I
Materials Research Foundations **37** (2018)

Materials Research Forum LLC
doi: http://dx.doi.org/10.21741/9781945291852

CHAPTER 1

Silicon Carbide Surface Cleaning and Etching

V. Jokubavicius*, M. Syväjärvi, R. Yakimova

Department of Physics, Chemistry and Biology (IFM), Linköping University, SE-58183, Linköping, Sweden

*valdas.jokubavicius@liu.se

Abstract

Silicon carbide (SiC) surface cleaning and etching (wet, electrochemical, thermal) are important technological processes in preparation of SiC wafers for crystal growth, defect analysis or device processing. While removal of organic, particulate and metallic contaminants by chemical cleaning is a routine process in research and industrial production, the etching which, in addition to structural defects analysis, can also be used to modify wafer surface structure, is very interesting for development of innovative device concepts. In this book chapter we review SiC chemical cleaning and etching procedures and present perspectives of SiC etching for new device development.

Keywords

Silicon Carbide, Chemical Cleaning, Wet Etching, Electrochemical Etching, Porous SiC

Contents

1. Introduction

Silicon carbide is a semiconductor material which has suitable properties for various electronic applications, mainly in medium- and high-voltage power devices [1,2]. Due to the excellent material characteristics SiC devices can outperform silicon counterparts in terms of blocking voltages and operating temperatures [3]. Electronic products incorporating SiC can be smaller, require less cooling and can operate under harsh conditions which are out of operational window for silicon-based technologies. For example, in hybrid/electrical vehicles, power devices based on SiC allow to improve fuel economy and offer higher flexibility in designing and positioning various system components [4–6]. Semiconductor companies have already introduced SiC based diodes and transistors into the market. However, SiC crystal growth and device processing have not reached similar maturity like silicon technologies. Intensive research and development is still being carried out regarding the single crystal SiC bulk growth process optimization and crystal diameter enlargement as well as on improvement of device processing procedures to reduce fabrication costs and increase device performance [6,7]. While industry is focusing on power technologies, academic research on SiC is shifting towards various innovative device concepts, some of which are discussed in this chapter.

In crystal growth and device processing, a clean SiC substrate surface is crucial in obtaining high yield and reliability. For example, solid particles present on the SiC substrate prior to epitaxial or bulk growth could lead to formation of structural defects which significantly deteriorate crystal quality and material yield. In device processing, improper surface cleanliness could cause numerous problems ranging from impair adhesion of photoresist films to formation of electrical defects in oxide layer. Basic chemical cleaning techniques for SiC were adapted from silicon industry. In contrast to chemical cleaning, wet etching for surface modification or structural defects analysis requires more aggressive process for SiC compared to silicon. In this book chapter, the most common SiC chemical cleaning, wet and thermal etching processes which are applied in research and development, are reviewed. Even though such processes have been explored for decades, there are still novel findings which could lead to new avenues in research and development. Some applications of wet and thermal etching for innovative SiC device concepts are also presented.

2.   Wet chemical cleaning of SiC

2.1   Surface contaminations

Most common contaminants on semiconductor surfaces are organic/molecular films, solid particles, various metals or their ions. In case of SiC, there is also a native oxide layer. It has been shown that silicon dioxide (SiO_2) with a thickness of 1 nm forms on SiC surface in a matter of minutes after chemical cleaning procedures [8,9]. Such oxide should be removed before measurements using surface sensitive techniques or growth of low dimensional materials such as graphene. In general, contaminants can originate from a wide range of sources such as the ambient, wafer dicing, chemicals, personnel etc. Organic compounds, which are present even in a clean-room air, readily adsorb on any semiconductor surface. They can form an organic film and mask some of the particles already residing on the surface. Upon high temperature growth of epitaxial layers or bulk material, organic contaminants carbonize and can form a nucleation point for various structural defects (microcracks, dislocations, stacking faults etc.). Similar defects in crystal growth will occur if solid particles are present on the surface. In device processing, solid particles would affect the entire device fabrication chain by being embedded in deposited films or acting as a mask in photolithographic processes. Metallic contaminants (Fe, Al, Ni, Cu etc.) and ionic metals (Ca, Na) from liquid chemicals, water, metallic tools used to handle SiC sample or measurement equipment could also influence semiconductor device fabrication. For example, commonly used mercury (Hg) probe capacitance-voltage measurement equipment can also leave traces of Hg on the SiC and have to be removed [10]. During heat treatment steps in device fabrication, metallic contaminants can diffuse into the semiconductor and introduce defect levels/traps in the band gap causing device degradation. Due to differences in materials parameters (chemical bond strength, diffusion coefficients, surface energies) of Si and SiC, the latter is less sensitive to diffusion of metallic contamination into bulk material at similar processing temperatures. However, it has been shown that some metallic impurities can degrade intrinsic lifetime of gate oxides [11,12]. Thus their removal and monitoring, for example using the total reflection X-ray fluorescence spectroscopy [13], are important issues in SiC MOSFET fabrication.

2.2   RCA, Piranha and HF cleaning

The most widely used and best-established cleaning solutions for SiC have been transferred from the Si industry. The most common chemistries are the well-known RCA (also called "standard clean" or "SC") and Piranha cleaning. Of course, before applying any of these cleaning procedures the sample can be pre-clean in ultrasonic bath using isopropanol at 80°C followed by rinsing in deionized water.

The RCA process was first introduced in device production at RCA (Radio Corporation of America) in 1965 and its detail description was published by W. Kern and D. D. Puotinen in 1970 [14]. At that time, it was a breakthrough in the semiconductor surface cleaning technologies and it still remains as the most widely used chemical cleaning procedure not only for Si, but for SiC as well. The process is based on a two-step oxidizing and complexing treatment with hydrogen peroxide solutions [15] :

- 1st step or SC1 - cleaning in an alkaline mixture (5 H$_2$0:1 H$_2$O$_2$(30%):1 NH$_4$OH(29%)). The composition of the solution can vary from 5:1:1 to 7:2:1 parts by volume of H$_2$O, H$_2$O$_2$, and NH$_4$OH. The treatment is done in ultrasonic bath for 5-10 min at solution temperature of 65-75 °C followed by rinsing in deionized (DI) water. The SC1 is mainly used for cleaning of organic contaminants which are removed by oxidative breakdown and dissolution. During the treatment the native oxide layer slowly dissolves and a new one is formed by oxidation. Such oxide regeneration process acts as a self-cleaning effect which aids to dislodge particles [16]. The SC1 also removes traces of IB, IIB metals (Au, Ag, Cu, Zn, Cd, Hg) and some other elements like Ni, Co, and Cr by complexing. The SC1 solution possesses poor thermal stability leading to decomposition of H$_2$O$_2$ to H$_2$O and oxygen, and evaporation of NH$_3$ from NH$_4$OH. Therefore, a freshly prepared mixture of chemicals has to be used for each cleaning cycle.

- 2nd step or SC2 – cleaning in an acid mixture (6 H$_2$0:1 H$_2$O$_2$(30%):1 HCl(27%)) using ultrasonic bath with solution temperature and treatment time similar to SC1. It removes alkali ions, and cations such as Al^{+3}, Fe^{+3} and Mg^{+2} as well as metals which were not completely removed during the first step. HCl in SC2 solution is used to increase oxidation strength and complexing of metals [17]. After the SC2 process the sample should be rinsed in DI water. Also, as in SC1, a freshly prepared mixture of chemicals is strongly recommended for each cleaning cycle.

Another strong acid wet cleaning used for SiC is a mixture of 4 H$_2$SO$_4$(98%): 1 H$_2$O$_2$(30%) so-called "Piranha" clean. The surface is exposed to this mixture for at least 10 min at 100-

130 °C. The Piranha clean is primarily used for removing the photoresist and heavy organic contaminants. The mixture is extremely dangerous to handle due to its great ability to eradicate organics. The advantage of Piranha over RCA cleaning for organic residues cleaning was demonstrated in SiC biocompatibility studies by S. Saddow et al. [18].

In the original RCA clean process for Si, an optional step of HF cleaning after SC1 can be introduced [14]. The main purpose of this step is to remove oxide film which may form during SC1 step and entrap some trace impurities. However, it was noted that such cleaning will result in recontamination of the Si surface if the HF solution is not of a very high purity and particle-free [16]. In addition, it was not advisable to use HF cleaning after SC2 step since it would cause loss of the protective SiO_2 film that passivates the silicon surface. In case of SiC, the HF treatment is used to remove SiO_2 before surface sensitive measurements or growth of low dimensional materials, such as graphene by sublimation [19], where presence of oxygen is not tolerated. The HF is typically diluted with DI water to slow down the etch rate of SiO_2 and obtain better etching uniformity. The dilution ratios can vary from 1 H_2O:1 HF to 100 H_2O:1 HF [20]. The etching time depends on dilution ratio. For example, the etch rate of SiO_2 with 10 H_2O:1 HF is ~10 Å/s [21,22]. The HF may be diluted with ammonium fluoride (NH_4F). In this case the treatment is called a Buffered Oxide Etch (BOE) or Buffered HF (BHF). Buffered HF is used instead of dilute HF when films, such as photoresist, that could be damaged by a highly acidic environment are present on the wafer surface or when there is a need to remove a large amount of oxide [17].

Numerous studies were performed to analyse wettability of differently processed SiC surfaces after HF cleaning [20,23,24]. King et al. [20] demonstrated that, in addition to chemical treatment, the wettability of SiC depends also on the initial state of the surface (Table 1).

Table 1. Wetting characteristics of as -polished and thermally oxidized 6H-SiC [20].

Treatment	6H-SiC (0001) as-polished	6H-SiC (0001) after oxide removal using HF
None	Hydrophobic	Hydrophilic
SC1	Hydrophilic	Hydrophilic
SC2	Hydrophobic	Hydrophilic
10 H_2O:1 HF	Hydrophobic	Hydrophilic
Piranha	Hydrophilic	Hydrophilic

The same authors showed that the hydrophilic nature of SiC surfaces produced in wet chemical cleaning could lead to trapping of contaminants in micropipes and their removal/outgassing in subsequent processing [25].

Despite the effectiveness of any cleaning technique the semiconductor surface can be contaminated again by improper rinsing, drying or post-cleaning storage. Cleaned glass or stainless-steel containers flushed with high purity nitrogen and stored in a clean room environment could be considered for sample storage [16]. For a longer storage, or transportation of SiC samples, evacuated plastic bags could also be used. Handling samples with metal tweezers would leave traces of metals on the surface. Using vacuum tweezers would at least allow avoiding such contamination on the front side of the sample.

3.   Chemical, electrochemical and thermal etching of SiC

3.1   Chemical etching

Silicon carbide is resistant to any chemical solutions at room temperature due to the very high chemical inertness [26]. Therefore, chemical etching of SiC can only be achieved by hot gases or molten salts and alkali solutions at high temperatures. Only etching using molten salts will be discussed in this chapter. The etching mechanism is based on breaking SiC bonds on the surface by reactive molecules from molten salts followed by formation of oxides which are subsequently dissolved in the same solution.

Various studies were dedicated to investigate the effect of different molten salts like $KClO_3$, KCl-NaCl, K_2CO_3, K_2SO_4, KNO_3, $Na_2B_4O_7$, Na_2CO_3, $NaNO_3$, Na_2SO_4, NaF: Na_2SO_4, PbF_2 and other, but in most cases there were problems with chemical attack to the crucible, need for high process temperature or stability of molten salt itself [26–35]. Today, the most common way to reveal structural defects in single crystal SiC wafers by chemical etching is to use molten KOH. A typical KOH etching set-up is made of ceramic heater equipped with thermocouple and nickel crucible. The thermocouple encapsulated in Ni tube can be dipped in molten KOH as close as possible to the sample to obtain more accurate temperature reading [36]. A small basket/holder made of nickel or platinum is used to immerse the SiC sample in molten KOH at 400–600 °C. The etching rates may vary depending on temperature, crystal orientation or etching atmosphere during the KOH etching. A very detailed study on etching rates of SiC in molten KOH has been done by Katsuno et al. [37]. The authors have reported that the etching rate at 520 °C is about four times higher on $(000\bar{1})$ surface ($\sim$2.3 µm/min) compared to (0001) surface ($\sim$0.6 µm/min) while etching rates of the $(11\bar{2}0)$ and the $(1\bar{1}00)$ surfaces are almost equal to that of the $(000\bar{1})$. In addition, they demonstrated that for n- and p-type samples the carrier concentration hardly influence the etching rate up to 3×10^{19} cm^{-3} and that the etching rate is enhanced by about 2% as the hexagonality of SiC crystals increases from the 6H(33%) < 15R(40%) < 4H(50%). Mokhov et al.

showed that the etching rate of the neutron-irradiated SiC polar faces in the KOH melt depends on the irradiation dose or density of radiation induced point defects [38].

The primary use of molten KOH is to reveal and analyze various defects on SiC surfaces. Crystal areas containing defects have higher strain energies compared to single crystal areas. Thus, they are more sensitive to chemical attack and this leads to preferential etching [39]. Various defects can be identified based on the shape of the etch pit and information on their densities and interactions can be obtained [40,41]. The most common defects in hexagonal SiC which can be revealed by molten KOH etching, are screw dislocations (SDs), threading edge dislocations (TEDs), basal plane dislocations (BPDs) [36,42–47] (Fig. 1a). Screw dislocations can be divided into closed-core (threading screw dislocations (TSDs)) and hollow-core (micropipes (MPs)) dislocations. The hollow cores become evident when the Burger vector exceeds lattice parameter at least two times in 6H SiC and three times in 4H-SiC [48,49]. Stacking faults occurring as linear etch pits after etching can be delineated on the cleavage of $(11\bar{2}0)$ and $(1\bar{1}00)$ planes [50–54]. Typical imaging and counting of defects can be performed using an optical microscope with Nomarski Differential Interface Contrast (NDIC). Sakwe et al. pointed out the importance of KOH etching process optimization in order to avoid over etching and merging of different etch pits [36]. There is a difference in etching selectivity on the (0001) and the $(000\bar{1})$ faces. Syväjärvi et al. showed that (0001) is etched preferentially, whereas the $(000\bar{1})$ is etched nearly isotropically [55]. Such anisotropy is clearly seen when comparing micropipe-related pit diameter on etching time on both polar faces of SiC (Fig. 1b). The openings related to micropipes are at least a factor of 10 smaller on the $(000\bar{1})$ face and they possess roundish shape instead of hexagonal on the (0001) face [55,56]. The shape of the etch pit can be greatly influenced by the conductivity and doping concentration. For example, the p-type (0001) face in 4H-SiC possesses much stronger preferential etching compared to n^+ samples, on which mainly round pits with continuous size distribution is formed [44]. The authors attributed this difference to band bending structure and injection of holes at the 4H-SiC/KOH interface depending on the conductivity type. Based on their theory, in the case of molten KOH etching of highly n-type 4H-SiC, electrochemical processes become dominant due to formation of a p-type inversion layer on the SiC surface and this is resulting in isotropic etching. Such dependence of molten KOH etching on doping made it very difficult to distinguish etch pits of TSDs and TEDs in heavily ($n>6\times10^{18}$ cm^{-3}) n-type doped SiC [44,60,61]. Nevertheless, it was shown that TSDs (or TEDs) can be identified with very high accuracy after etching in molten KOH:Na$_2$O$_2$=50:3 [61]. Another report showed that the etch pit size of TSDs and TEDs in highly n-type 4H-SiC differs more greatly when more than 20-wt.% Na$_2$O$_2$ is added to the KOH [62]. The authors claimed that such improvement was achieved due to dissolved oxygen which enhanced the defect-selective anisotropic etching. Cui et al reported that combination of molten KOH etching with laser confocal microscopy can also be a powerful tool to identify and characterize TEDs and TSDs in n-type and semi-insulating SiC wafers based on etch pit sectional view information and etch pit angle [38].

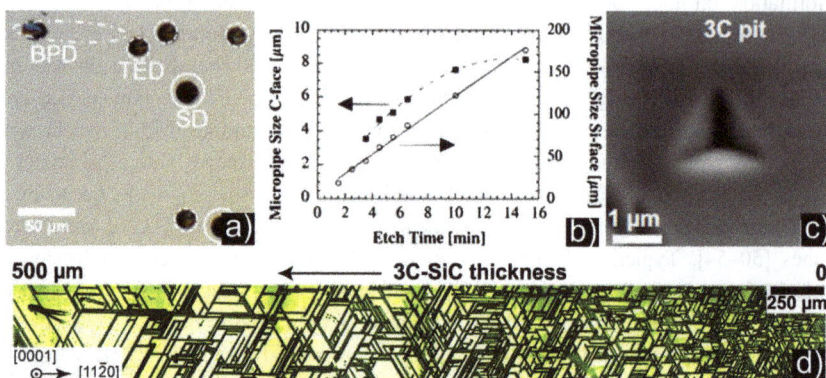

Figure 1. a) etch pits in 4H-SiC (0001) [57] (Copyright 2014 The Japan Society of Applied Physics),
b) micropipe related pit diameter vs. etching time for Si- and C-faces [55] (Reproduced with permission from J. Electrochem. Soc., 147, 3519 (2000). Copyright 2000, The Electrochemical Society.),
c) etch pit in 3C-SiC(111) after etching in molten KOH [58] (P.G. Neudeck, A.J. Trunek, D.J. Spry, J.A. Powell, H. Du, M. Skowronski, X.R. Huang, M. Dudley, CVD Growth of 3C-SiC on 4H/6H Mesas, Chem. Vap. Depos. 2006, 12, 531–540. Copyright Wiley-VCH Verlag GmbH & Co. KGaA. Reproduced with permission.),
d) stacking faults density vs thickness in 3C-SiC(111) revealed by etching in KOH [59] (Reprinted with permission from Cryst. Growth & Des. 15, 2940-2947 © (2015) American Chemical Society).

Defects which can be revealed by KOH etching in cubic SiC are stacking faults (SFs), double positioning boundaries (DPBs) and threading edge dislocations (TEDs). Triangular etch pit (Fig. 1c) is a characteristic feature of threading screw dislocations in (111) crystals [39]. Stacking faults density variation with the thickness can be analyzed in 3C-SiC (111) grown on off-oriented hexagonal SiC substrates (Fig. 1d) [59].

3.2   Electrochemical etching

Unlike chemical etching described in Section 3.1 the electrochemical etching can be done at room temperature. In the electrochemical etching cell, the SiC sample and a counter electrode are immersed in electrolyte. By applying an external voltage between them, one can regulate the electric field in the space charge region to inject holes at the interface with the electrolyte what causes oxidation and dissolution of SiC material.

In case of p-type material, the etching process can occur in darkness while etching of n-type material needs UV illumination to generate minority carriers [63]. The most common electrolytes used in electrochemical etching are fluoride or alkaline based solutions: HF, KOH, NaOH, H_2SO_4, HCl, and H_2O_2 [27]. The vast majority of studies in electrochemical etching of SiC have been done using HF. It has been suggested that the following reactions occur during the electrochemical etching of SiC in HF solutions [64,65]:

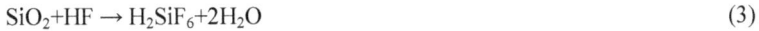

$$SiC+4H_2O+8h^+ \rightarrow SiO_2+CO_2+8H^+ \tag{1}$$

$$SiC+2H_2O+4h^+ \rightarrow SiO+CO+4H^+ \tag{2}$$

$$SiO_2+HF \rightarrow H_2SiF_6+2H_2O \tag{3}$$

The first two equations describe electrochemical oxidation while the third equation shows dissolution of oxide in HF solution. Electrochemical etching of SiC leads to formation of porous structure. There are always nanometer size fibers left behind as the etching front moves into the bulk material. The protection of porous structures from being completely etched away as the porosity increases was explained by formation of space charge region at the SiC/electrolyte interface [66,67]. The morphology of the etched SiC depends on polarity [68,69], current density [70] and electrolyte solution [71]. It has been shown that n-type heavily doped 4H-SiC wafers can be electrochemically anodized in a HF-based electrolyte without any UV light assistance [68]. The authors used HF mixed with acetic acid (to increase electrolyte wettability) and water with volume ratios HF(50%):acetic acid:H_2O of 4.6:2.1:1.5. Usually a dendritic porous structure is formed on (0001) face while (000$\bar{1}$) face leads to columnar structure [68,69]. Also, it was reported that a thin skin layer on top of the porous structure can form inducing an inhomogeneous pattern of the pores [69,72]. Inhomogeneities in pore structures were obtained by applying constant current or constant voltage conditions [67,73]. In contrast, porous 4H-SiC with tailored properties can be prepared by combining metal assisted photochemical etching (MAPCE) and photoelectrochemical etching (PECE) steps [74–76]. A noble metal is deposited on SiC surface in MAPCE process. The sample is then immersed in HF solution containing

Advancing Silicon Carbide Electronics Technology I   Materials Research Forum LLC
Materials Research Foundations 37 (2018)   doi: http://dx.doi.org/10.21741/9781945291852

oxidizing agent. The deposited metal acts as local cathode where the oxidizing agent is reduced, and the surface areas not covered with the metal get porosified. A porous layer generated with MAPCE provides initiation sites for PECE, thus the formation of a skin layer can be avoided.

3.3   Thermal etching

Various SiC surface contaminants can be removed using chemical cleaning procedures such as those described in section 3.1. However, removal of surface scratches remaining after mechanical polishing or modification of surface steps structure require more reactive surface treatment such as hydrogen etching or chemo-mechanical polishing (CMP). Hydrogen etching is performed *in situ* in the CVD reactor using fluxes of H_2+HCl, $H_2+C_3H_8$ or H_2+SiH_4 gases at temperatures of 1300-1600 °C [77,78]. In addition to removal of surface scratches, it is possible to control surface step morphology, but it is quite an expensive process. The CMP combines mechanical polishing and chemical etching simultaneously using special slurries [79–81], thus very smooth surfaces can be obtained. However, surface structure engineering is not possible, since the CMP is suitable only for surface planarization. An alternative technique, which does not require any aqueous solutions or dangerous gases, is thermal etching. Originally it has been used to pre-etch SiC seeds by reversing temperature gradient during sublimation growth of bulk SiC crystals [82]. Since the quality of the SiC seeds has significantly improved since 90s, the thermal etching is mainly used for SiC surface step engineering. Thermal etching of SiC is based on sublimation decomposition of SiC at elevated temperatures [83–87]. The etching of SiC surface is usually performed in a graphite crucible. Upon sublimation the SiC surface decomposes with the highest partial vapor pressure of Si followed by SiC_2 and Si_2C [88]. The composition of vapor species above the SiC crystal depends on temperature and it has been calculated that the atomic flow of C is less than 1% of the atomic flow of silicon in the Si-C chemical system [88]. Therefore, if loss of Si is not compensated and accumulation of C on the surface is not prevented, there is graphitization of SiC surface upon high temperature annealing. It has been demonstrated that graphitization can be significantly suppressed by placing Ta near the SiC surface [89]. In the Si-C-Ta material system Ta reacts with the carbon containing vapor species by forming stable carbide. Therefore, in the etching cavity with Ta, the following reaction is thought to occur:

$$Ta\ (solid) + Si_2C(gas) \rightarrow TaC(solid) + 2Si(gas) \qquad (4)$$

This implies that Ta creates more Si-rich environment, Si reacts with carbon bearing species and provides conditions for thermal etching of SiC surface. Lebedev et al. demonstrated that atomically flat 6H-SiC (0001) surface with terraces separated by steps of unit-cell height (h = 1.5 nm) can be obtained on mechanically polished 6H-SiC (0001) after etching in vacuum ($\sim 10^{-6}$ mbar) at 1300-1400 °C [83]. Nishiguchi et al. investigated effect of nitrogen and argon ambient to the etching of ($11\bar{2}0$), ($000\bar{1}$) and (0001) surfaces of 6H-SiC at 2500 °C [84]. They found that etching in nitrogen atmosphere provides better surface morphology compared to argon. In addition, step bunching was a serious problem in obtaining flat surfaces on 6H-SiC (0001) substrates, but it hardly occurred on the ($11\bar{2}0$) substrates and atomically flat surface was obtained (Fig. 2a-f). Jokubavicius et al. performed thermal etching experiments of 6H-SiC (0001), 4H-SiC (0001) and 3C-SiC (111) surfaces in argon (700 mbar) and vacuum ($\sim 10^{-5}$ mbar) at isothermal conditions (Fig. 2g) [87]. They observed that after thermal etching the smoothest surfaces can be obtained in 3C-SiC(111) and attributed that to differences in step bunching formation in hexagonal and cubic polytypes. All SiC polytypes are composed of SiC bilayers which are stacked on top of each other. The energies of interaction between each SiC bilayer plane vary depending on unique stacking sequence of specific polytype. This induces variations in step dynamics and formation of step bunching upon crystal growth or sublimation/thermal etching. The 3C-SiC is the only polytype in which the interaction energy between different SiC bilayer planes is the same, thus energetically driven step bunching is not expected [90,91].

4. Perspectives of SiC etching for various device concepts

4.1 Fluorescent SiC for white LEDs

In most cases, wet chemical etching is used to fabricate porous structure on single crystal SiC surface. Considering small size of the pores one can expect quantum size effects to occur upon light illumination and this is primarily interesting for optoelectronic applications. Matsumoto et al. showed that blue-green emission from n-type 6H-SiC can be tuned by porous structure electrochemically etched in HF-ethanol solution [92]. The blue shift in the electroluminescence of n-type 6H-SiC with porous structure etched in HF solution illuminated with UV light was demonstrated by Rittenhouse et al. [93]. The properties of porous structures have been realized to be very useful in white light emission diode (LED) based on fluorescent SiC layer which converts UV to visible light. The initial idea of such white LED concept based on fluorescent SiC was proposed by Kamiyama et al. [94,95]. The proposed structure of the device contained donor and acceptor doped SiC as a phosphor material, called fluorescent SiC, which is excited by

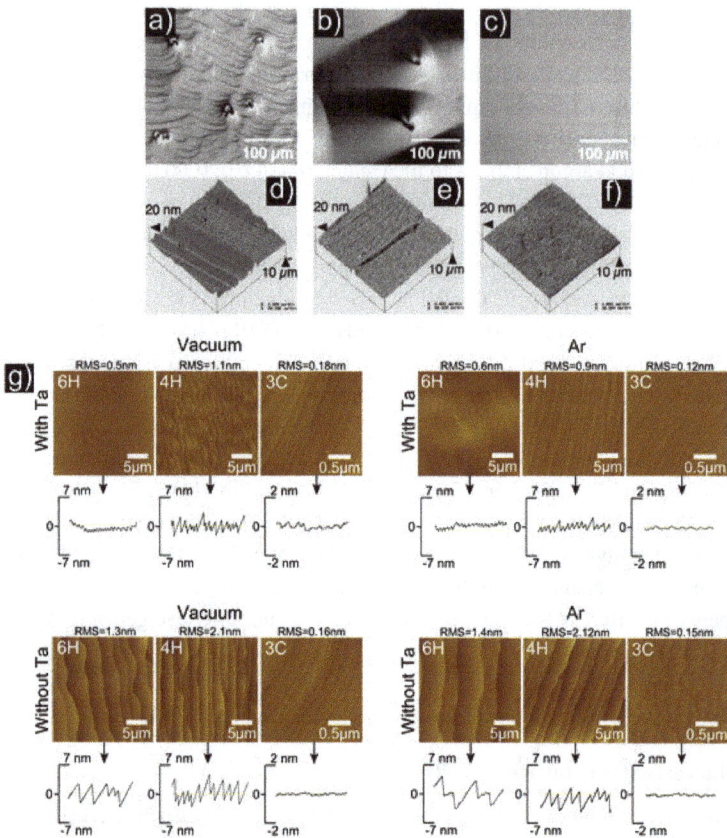

Figure 2. Optical images (a, b, c) and AFM images (d,e,f) of 6H-SiC substrates after thermal etching at T=2500 °C and p=700 Torr in nitrogen atmosphere for 30 min. (a and d) Si-face, (b and e) C-face, (c and f) (11$\overline{2}$0) surface (Copyright 2003 The Japan Society of Applied Physics) [84]. g) AFM images of 6H-, 4H-SiC (0001) and 3C-SiC(111) after etching for 10 min at 1800°C using different arrangements and ambient conditions[87] (Reprinted with permission from V. Jokubavicius, G.R. Yazdi, I.G. Ivanov, Y. Niu, A. Zakharov, T. Iakimov, M. Syväjärvi, R. Yakimova, Surface engineering of SiC via sublimation etching, Appl. Surf. Sci. 390 (2016) 816. Copyright (2016) Elsevier B.V.).

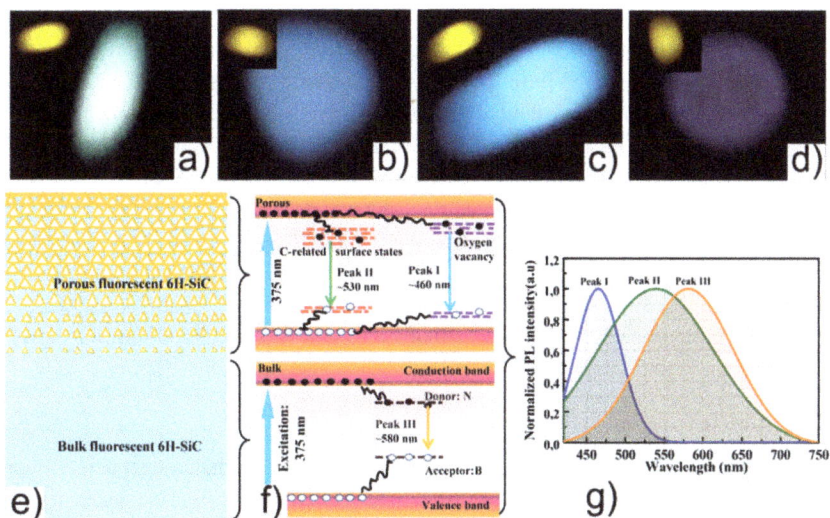

Figure 3. PL images of porous fluorescent SiC fabricated by electrochemical etching in a) diluted HF, emission peak at 491 nm b) diluted HF+$K_2S_2O_8$ (0.01 mol), emission peak at 449 nm, c) diluted HF+ $K_2S_2O_8$ (0.015 mol), emission peak at 434 nm, d) diluted HF+ $K_2S_2O_8$ (0.02 mol), emission peak at 407nm. Yellow insets in a-d correspond to emission from N-B doped 6H-SiC at about 580 nm [96] (T. Nishimura, K. Miyoshi, F. Teramae, M. Iwaya, S. Kamiyama, H. Amano, I. Akasaki, High efficiency violet to blue light emission in porous SiC produced by anodic method, Phys. Status Solidi. 7 (2010) 2459– 2462. Copyright Wiley-VCH Verlag GmbH & Co. KGaA. Reproduced with permission.), e) schematic diagram of the fluorescent 6H-SiC with porous surface layer, f) three possible transitions related to: I) surface defects in porous SiC, II) oxygen vacancy, III) DAP recombination in N-B doped 6H-SiC, g) emission spectra from porous SiC layer [97].

UV light produced by nitride LEDs grown on doped SiC. Donor and acceptor pairs (DAPs) of N-B and N-Al in 6H-SiC layers can cover almost the entire visible spectral range. Such fluorescent SiC would possess uniform and stable color quality and an excellent thermal conductivity for high-power-operated LEDs. Moreover, SiC is well established substrate for nitride growth, thus a monolithic LED structure containing nitrides and fluorescent SiC is possible to fabricate. However, the emission efficiency of N-Al DAPs is much lower than that of N-B DAPs due to much higher thermal ionization

Advancing Silicon Carbide Electronics Technology I   Materials Research Forum LLC
Materials Research Foundations **37** (2018)   doi: http://dx.doi.org/10.21741/9781945291852

for shallower Al-acceptor states. Therefore, the emission from N-Al should be obtained by other means. Nishimura et al demonstrated that this could be done by etching porous structure on N-B doped SiC [96]. To control the porosity of N-B doped SiC they used electrochemical etching in HF diluted with $K_2S_2O_8$. This gives effect on light emission as shown in Fig. 3a-d. The latter study was expanded by Lu et al who introduced a passivation layer of Al_2O_3 on porous structure to significantly enhance the emission of light (Fig. 3e-g) [97]. Moreover, the authors demonstrated that porous layers fabricated in commercial n-type and lab grown N-B co-doped 6H-SiC show emission peaks centered approximately at 460 nm and 530 nm. Those peaks were attributed to neutral oxygen vacancies and C-related surface defects generated during anodic oxidation process. The B-N DAP emission peak in 6H-SiC is usually positioned at about 580 nm. As shown in Fig. 3g, all three peaks can cover very broad visible light spectrum.

4.2   Rugate mirrors

As the porosity of etched SiC layer increases and structural inhomogeneities become smaller than typical wavelengths, the porous layer possesses refractive index n which depends on the structure. Higher porosity typically leads to a lower refractive index since the dielectric effective medium contains more air. Such engineering of optical properties in porous SiC can be used to fabricate rugate mirrors. The advantage of SiC based rugate mirrors is the possibility to use them in high temperature and chemically aggressive environment which is too hostile for other materials. As mentioned in section 3.2 of this book chapter, the conventional electrochemical etching of SiC can lead to formation of a skin layer on top of the porous structure or an inhomogeneous pattern of the pores. These problem can be solved by combining metal assisted photochemical etching (MAPCE) and photoelectrochemical etching (PECE) steps [74,75].

The fabrication of rugate mirror is divided in three steps:

1) To avoid skin and cap layer formation a porous SiC layer with the thickness of about 1 µm is formed by MAPCE.

2) PECE with etching solution (0.04M $Na_2S_2O_8$ in 1.31 mol/l HF) exposed to UV light is used to fabricated SiC layer with uniform porosity with the thickness of about 30 µm by varying the voltage as a function of transferred charge [98].

3) By applying two 60 V pulses, each lasting 6 s, at the end of the etching process the porous layer is separated from the bulk material which can be re-used for another mirror fabrication (Fig. 4b).

Materials Research Forum LLC

doi: http://dx.doi.org/10.21741/9781945291852

Such mirrors exhibit reflectance peaks at 375–385 nm and 710–750 nm (Fig. 4c). The peaks can be shifted depending on the etching conditions resulting in different porous SiC structure and refractive index.

Figure 4. a) Cross sectional SEM micrograph of porous SiC rugate mirror [98], b) detached porous 4H-SiC layer after PECE (on the right) and corresponding substrate (on the left) [75], c) three reflectance measurements at different positions indicated by different colors [98].

4.4 Porous SiC membranes for biomedical applications

Porous SiC layers could be also used as semipermeable membranes for microdialysis probes [18]. It has been demonstrated that porous SiC allows diffusion of proteins, discourages biofouling by high protein solutions and clot formation in the bloodstream [99,100]. A free standing n- and p-type porous SiC membranes were fabricated using a three-electrode arrangement where a 6H-SiC crystal serves as the working electrode, a saturated calomel electrode as the reference, and a platinum disk as the counterelectrode [100]. At the beginning the surface of 6H-SiC was photoelectrochemically etched (p-type samples in dark, n-type samples under UV light) in HF + ethanol solution. Then, the porous films exhibiting dendritic structure were separated by applying a high current density (up to 100 mA/cm^2). In additional, authors also fabricated n-type porous SiC with a columnar structure using two-electrode electrochemical cell with a SiC sample as the anode and a platinum plate as the cathode [99]. Based on their results, porous SiC has a low protein adsorption comparable to the best commercially available polymeric membranes what makes it very promising for medical microsystem applications.

4.5 Graphene nanoribbons

Low dimensional graphitic structures, like graphene nanoribbons, exhibit exceptional electronic transport properties that make them attractive for nanoscale devices [101,102]. It has been shown that electrons can travel more than 10 μm in graphene nanoribbons without scattering [103]. However, the challenge is to develop a reliable nanoribbons fabrication technique since patterning graphene sheets using conventional lithography introduces defects at edges while growth of graphene nanoribbons on metallic substrates requires graphene transfer processes. Graphene growth on the sidewalls of trenches etched in the (0001) surface of semi-insulating SiC substrate is one of most promising ways to fabricate graphene nanoribbons [103,105]. One of the most important steps before formation of graphene nanoribbons is thermal etching or annealing process of pre-patterned SiC for formation of facets on the sidewalls on which growth of epitaxial graphene nanoribbons takes place (Fig. 5a-c).

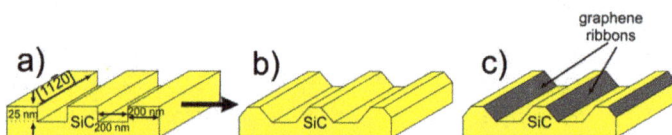

Figure 5. a) Structured SiC surface using lithography, b) formation of facets after thermal etching/annealing at 900°C for 30 min, c) preferential growth of graphene on the sidewalls, which would lead to isolated graphene nanoribbons on the facets [104]*(A. Stöhr, J. Baringhaus, J. Aprojanz, S. Link, C. Tegenkamp, Y. Niu, A.A. Zakharov, C. Chen, J. Avila, M.C. Asensio, others, Graphene Ribbon Growth on Structured Silicon Carbide, Ann. Phys. 529 (2017) 1700052. Copyright Wiley-VCH Verlag GmbH & Co. KGaA. Reproduced with permission.).*

Stöhr et al processed trenches of 25 nanometer in depth and periodicity of 400 nm [104]. The authors broke down the process in two steps to form graphene nanoribbons. In the first one, the sample was annealed or thermally etched in UHV at 1200 °C for 30 min to enable a sufficient SiC mass transport, which leads to thermal etching of the sidewalls into facets. In the second step, the annealing temperature was increased to around 1400 °C for 10 minutes to initiate selective growth of graphene nanoribbons. In both steps, so called a "face to face" annealing configuration which is obtained by placing two SiC substrates one on top of the other face to face, with a small gap in between, and then simultaneous heating is used [106]. At high temperatures such configuration provides similar effects as placing Ta in close proximity to the SiC surface. In both cases graphitization of SiC surface is significantly suppressed. In the first case the loss of silicon is compensated by the sublimation of counter SiC surface, while in the second case, tantalum reacts with carbon bearing species what makes the stoichiometry above the SiC surface enriched with silicon.

5. Summary

In this book chapter the most common SiC chemical cleaning (RCA, Piranha), wet and thermal etching techniques for defects analysis and surface/bulk crystal structure modifications have been reviewed. In addition, some innovative (non-power) device applications like fluorescent SiC, rugate mirrors, graphene nanoribbons and SiC membranes have been presented.

References

[1]   X. She, A.Q. Huang, O. Lucia, B. Ozpineci, Review of Silicon Carbide Power Devices and Their Applications, IEEE Trans. Ind. Electron. 64 (2017) 8193–8205. https://doi.org/10.1109/TIE.2017.2652401

[2]   A. Morya, M. Moosavi, M.C. Gardner, H.A. Toliyat, Applications of Wide Bandgap (WBG) devices in AC electric drives: A technology status review, in: 2017 IEEE Int. Electr. Mach. Drives Conf., 2017. https://doi.org/10.1109/IEMDC.2017.8002288

[3]   A. Elasser, T.P. Chow, Silicon carbide benefits and advantages for power electronics circuits and systems, Proc. IEEE. 90 (2002) 969–986. https://doi.org/10.1109/JPROC.2002.1021562

[4]   B. Whitaker, A. Barkley, Z. Cole, B. Passmore, D. Martin, T.R. McNutt, A.B. Lostetter, J.S. Lee, K. Shiozaki, A high-density, high-efficiency, isolated on-board vehicle battery charger utilizing silicon carbide power devices, IEEE Trans. Power Electron. 29 (2014) 2606–2617. https://doi.org/10.1109/TPEL.2013.2279950

[5]   F. Wada, N. Miyamoto, K. Yoshida, S. Godo, 6-in-1 Silicon Carbide (SiC) MOSFET Power Module for EV/HEV inverters, in: PCIM Asia 2017; Int. Exhib. Conf. Power Electron. Intell. Motion, Renew. Energy Energy Manag., 2017: pp. 1–4.

[6]   M. Su, C. Chen, S. Sharma, J. Kikuchi, Performance and cost considerations for SiC-based HEV traction inverter systems, in: 2015 IEEE 3rd Work. Wide Bandgap Power Devices Appl., 2015: pp. 347–350. https://doi.org/10.1109/WiPDA.2015.7369032

[7]   A.R. Powell, J.J. Sumakeris, Y. Khlebnikov, M.J. Paisley, R.T. Leonard, E. Deyneka, S. Gangwal, J. Ambati, V. Tsevtkov, J. Seaman, others, Bulk Growth of Large Area SiC Crystals, in: Mater. Sci. Forum, 2016: pp. 5–10.

[8]   Z. Zolnaia, N.Q. Khánha, E. Szilágyib, Z.E. Horvátha, T. Lohnera, Native oxide
      and ion implantation damaged layers on silicon carbide studied by ion beam
      analysis and ellipsometry, in: Proc." XV Int. Conf. Phys. Students ICPS, 2000: pp.
      4–11.

[9]   W. Huang, X. Liu, X.C. Liu, T.Y. Zhou, S.Y. Zhuo, Y.Q. Zheng, J.H. Yang, E.W.
      Shi, Nano-Scale Native Oxide on 6H-SiC Surface and its Effect on the Ni/Native
      Oxide/SiC Interface Band Bending, Mater. Sci. Forum. (2014).
      https://doi.org/10.4028/www.scientific.net/MSF.778-780.566

[10] J.J. McMahon, M. Jahanbani, S. Arthur, D. Lilienfeld, P. Gipp, T. Gorczyca, J.
      Formica, L. Shen, M. Yamagami, B. Hillard, J. Byrnes, Wet Processing for Post-
      epi & Pre-furnace Cleans in Silicon Carbide Power MOSFET Fabrication, ECS
      Trans. 69 (2015) 269–276. https://doi.org/10.1149/06908.0269ecst

[11] K.F. Schuegraf, C. Hu, Reliability of thin SiO2, Semicond. Sci. Technol. (1994).

[12] B.D. Choi, D.K. Schroder, Degradation of ultrathin oxides by iron contamination,
      Appl. Phys. Lett. (2001). https://doi.org/10.1063/1.1410363

[13] H. Kohno, Evaluation of Contamination of Power Semiconductor Device Wafers
      by TXRF Spectrometer, Rikagu J. 29 (2013) 9–14.

[14] W. Kern, D. Puotinen, Cleaning Solutions Based on Hydrogen for Use in Silicon
      Semiconductor Technology, R.C.A. Rev. 31 (1970) 187–206.

[15] W. Kern, The Evolution of Silicon Wafer Cleaning Technology, J. Electrochem.
      Soc. 137 (1990) 1887–1892. https://doi.org/10.1149/1.2086825

[16] W. Kern, Handbook of semiconductor wafer cleaning technology, 1993.

[17] Y. Nishi, R. Doering, Handbook of Semiconductor Manufacturing Technology,
      Second Edition, CRC Press, 2017.

[18] S. Saddow, Silicon Carbide Biotechnology: A Biocompatible Semiconductor for
      Advanced Biomedical Devices and Applications, Elsevier Science, 2016.

[19] C. Virojanadara, M. Syv\ajarvi, R. Yakimova, L. Johansson, A. Zakharov, T.
      Balasubramanian, Homogeneous large-area graphene layer growth on 6H-
      SiC(0001), Phys.Rev.B. 78 (2008) 245403.
      https://doi.org/10.1103/PhysRevB.78.245403

[20] S.W. King, R.J. Nemanich, R.F. Davisa, Wet Chemical Processing of (0001)Si
      6H-SiC Hydrophobic and Hydrophilic Surfaces, J. Electrochem. Soc. 146 (1999)
      1910–1917. https://doi.org/10.1149/1.1391864

[21] J.S. Judge, A Study of the Dissolution of SiO2 in Acidic Fluoride Solutions, J.
      Electrochem. Soc. . 118 (1971) 1772–1775. https://doi.org/10.1149/1.2407835

[22]   S. Verhaverbeke, The Etching Mechanisms of SiO2 in Hydrofluoric Acid, J. Electrochem. Soc. (1994). https://doi.org/10.1149/1.2059243

[23]   V. Stambouli, D. Chaussende, M. Anikin, G. Berthomé, V. Thoreau, J.C. Joud, Wettability Study of SiC in Correlation with XPS Analysis, in: Silicon Carbide Relat. Mater. 2003, Trans Tech Publications, 2004: pp. 423–426.

[24]   R.P. Socha, K. Laajalehto, P. Nowak, Influence of the surface properties of silicon carbide on the process of SiC particles codeposition with nickel, Colloids Surfaces A Physicochem. Eng. Asp. 208 (2002) 267–275. https://doi.org/10.1016/S0927-7757(02)00153-X

[25]   S.W. King, M.C. Benjamin, R.S. Kern, R.J. Nemanich, R.F. Davis, Ex Situ and In Situ Methods for Complete Oxygen and Non-Carbidic Carbon Removal from (0001)SI 6H-SiC Surfaces, MRS Proc. 423 (1996) 563.

[26]   V.J. Jennings, The etching of silicon carbide, in: Silicon Carbide–1968, Elsevier, 1969: pp. S199–S210. https://doi.org/10.1016/B978-0-08-006768-1.50023-1

[27]   D. Zhuang, J.H. Edgar, Wet etching of GaN, AlN, and SiC: A review, Mater. Sci. Eng. R Reports. 48 (2005) 1–46. https://doi.org/10.1016/j.mser.2004.11.002

[28]   T. Nakagawa, M. Hara, K. Imai, Hot corrosion behavior of SiC in molten Na_2SO_4, Nippon Kinzoku Gakkaishi/Journal Japan Inst. Met. 61 (1997) 1241–1248. https://doi.org/10.2320/jinstmet1952.61.11_1241

[29]   T. Sato, Y. Kanno, M. Shimada, Corrosion of SiC, Si_3N_4 and AIN in molten $K_2SO_4K_2CO_3$ salts, Int. J. High Technol. Ceram. 2 (1986) 279–290. https://doi.org/10.1016/0267-3762(86)90021-4

[30]   N.S. Jacobson, J.L. Smialek, Molten salt corrosion of alpha -SiC, in: Electrochem. Soc. Ext. Abstr., 1985: pp. 550–551.

[31]   R.E. Tressler, M.D. Meiser, T. Yonushonis, Molten Salt Corrosion of SiC and Si_3N_4 Ceramics-, J. Am. Ceram. Soc. 59 (1976) 278–279. https://doi.org/10.1111/j.1151-2916.1976.tb10962.x

[32]   J.W. Faust Jr, Processing of Silicon Carbide for Devices, in: Silicon Carbide High Temp. Semicond. Proc. Conf. Silicon Carbide, Boston, Mass. April. 1959, 1960: p. 403.

[33]   T. Gabor, V.J. Jennings, Effect of stirring on etching characteristics of silicon carbide, Electrochem. Technol. 3 (1965) 31.

[34]   S. Amelinckx, G. Strumane, W.W. Webb, Dislocations in Silicon Carbide, J. Appl. Phys. 31 (1960) 1359–1370. https://doi.org/10.1063/1.1735843

[35]   G.L. Harris, Properties of Silicon Carbide, INSPEC, Institution of Electrical Engineers, 1995.

[36]   S.A. Sakwe, R. Müller, P.J. Wellmann, Optimization of KOH etching parameters for quantitative defect recognition in n- and p-type doped SiC, J. Cryst. Growth. 289 (2006) 520–526. https://doi.org/10.1016/j.jcrysgro.2005.11.096

[37]   M. Katsuno, N. Ohtani, J. Takahashi, H. Yashiro, M. Kanaya, Mechanism of molten KOH etching of SiC single crystals: Comparative study with thermal oxidation, Japanese J. Appl. Physics, Part 1 Regul. Pap. Short Notes Rev. Pap. (1999).

[38]   Y. Cui, X. Hu, X. Xie, X. Xu, Threading dislocation classification for 4H-SiC substrates using the KOH etching method, CrystEngComm. 20 (2018) 978–982. https://doi.org/10.1039/C7CE01855J

[39]   K. Sangwal, Etching of crystals: Theory, Exp. Appl. (1987).

[40]   S. Ha, H.J. Chung, N.T. Nuhfer, M. Skowronski, Dislocation nucleation in 4H silicon carbide epitaxy, J. Cryst. Growth. 262 (2004) 130–138. https://doi.org/10.1016/j.jcrysgro.2003.09.054

[41]   I. Kamata, H. Tsuchida, T. Jikimoto, K. Izumi, Structural transformation of screw dislocations via thick 4H-SiC epitaxial growth, Jpn. J. Appl. Phys. 39 (2000) 6496–6500. https://doi.org/10.1143/JJAP.39.6496

[42]   J. Hassan, A. Henry, P.J. McNally, J.P. Bergman, Characterization of the carrot defect in 4H-SiC epitaxial layers, J. Cryst. Growth. 312 (2010) 1828–1837. https://doi.org/10.1016/j.jcrysgro.2010.02.037

[43]   S. Mahajan, M. V. Rokade, S.T. Ali, K. Srinivasa Rao, N.R. Munirathnam, T.L. Prakash, D.P. Amalnerkar, Investigation of micropipe and defects in molten KOH etching of 6H n-silicon carbide (SiC) single crystal, Mater. Lett. 101 (2013) 72–75. https://doi.org/10.1016/j.matlet.2013.03.079

[44]   Y. Gao, Z. Zhang, R. Bondokov, S. Soloviev, T. Sudarshan, The Effect of Doping Concentration and Conductivity Type on Preferential Etching of 4H-SiC by Molten KOH, in: Mater. Res. Soc. Symp. Proc., 2004: p. 815.

[45]   H. Wang, S. Sun, M. Dudley, S. Byrappa, F. Wu, B. Raghothamachar, G. Chung, E.K. Sanchez, S.G. Mueller, D. Hansen, M.J. Loboda, Quantitative comparison between dislocation densities in offcut 4H-SiC wafers measured using synchrotron X-ray topography and molten KOH etching, J. Electron. Mater. 42 (2013) 794–798. https://doi.org/10.1007/s11664-013-2527-x

[46]   K.M. Speer, P.G. Neudeck, D.J. Spry, A.J. Trunek, P. Pirouz, Cross-sectional
       TEM and KOH-Etch studies of extended defects in 3C-SiC p^+n junction diodes
       grown on 4H-SiC mesas, J. Electron. Mater. 37 (2008) 672–680.
       https://doi.org/10.1007/s11664-007-0297-z

[47]   J.L. Weyher, S. Lazar, J. Borysiuk, J. Pernot, Defect-selective etching of SiC,
       Phys. Status Solidi Appl. Mater. Sci. 202 (2005) 578–583.
       https://doi.org/10.1002/pssa.200460432

[48]   W. Si, M. Dudley, R.C. Glass, C.H. Carter Jr., V.F. Tsvetkov, Experimental
       Studies of Hollow-Core Screw Dislocations in 6H-SiC and 4H-SiC Single
       Crystals, in: Silicon Carbide, III-Nitrides Relat. Mater., Trans Tech Publications,
       1997: pp. 429–432.

[49]   M. Dudley, W. Si, S. Wang, C. Carter, R. Glass, V. Tsvetkov, Quantitative
       analysis of screw dislocations in 6H- SiC single crystals, Nuovo Cim. D. 19
       (1997) 153–164. https://doi.org/10.1007/BF03040968

[50]   N. Ohtani, M. Katsuno, T. Fujimoto, Reduction of stacking fault density during
       SiC bulk crystal growth in the [11$\bar{2}$0] direction, Jpn. J. Appl. Phys. 42 (2003) L
       277–L 279.

[51]   M. Syväjärvi, R. Yakimova, E. Janzén, Cross-sectional cleavages of SiC for
       evaluation of epitaxial layers, J. Cryst. Growth. 208 (2000) 409–415.
       https://doi.org/10.1016/S0022-0248(99)00484-4

[52]   T. Ohshima, K.K. Lee, Y. Ishida, K. Kojima, Y. Tanaka, T. Takahashi, M.
       Yoshikawa, H. Okumura, K. Arai, T. Kamiya, The electrical characteristics of
       metal-oxide-semiconductor field effect transistors fabricated on cubic silicon
       carbide, Japanese J. Appl. Physics, Part 2 Lett. 42 (2003).

[53]   J. Takahashi, N. Ohtani, M. Kanaya, Structural defects in α-SiC single crystals
       grown by the modified-Lely method, J. Cryst. Growth. 167 (1996) 596–606.
       https://doi.org/10.1016/0022-0248(96)00300-4

[54]   J. Takahashi, N. Ohtani, M. Katsuno, S. Shinoyama, Sublimation growth of 6H-
       and 4H-SiC single crystals in the [1$\bar{1}$00] and [11$\bar{2}$0] directions, J. Cryst. Growth.
       181 (1997) 229–240. https://doi.org/10.1016/S0022-0248(97)00289-3

[55]   M. Syväjärvi, R. Yakimova, E. Janzen, Anisotropic Etching of SiC, J.
       Electrochem. Soc. 147 (2000) 3519–3522. https://doi.org/10.1149/1.1393930

[56]   R. Yakimova, A.-L. Hylén, M. Tuominen, M. Syväjärvi, E. Janzen, Preferential
       etching of SiC crystals, Diam. Relat. Mater. 6 (1997) 1456–1458.
       https://doi.org/10.1016/S0925-9635(97)00076-9

[57] C. Kawahara, J. Suda, T. Kimoto, Identification of dislocations in 4H-SiC epitaxial layers and substrates using photoluminescence imaging, Jpn. J. Appl. Phys. 53 (2014) 20304. https://doi.org/10.7567/JJAP.53.020304

[58] P.G. Neudeck, A.J. Trunek, D.J. Spry, J.A. Powell, H. Du, M. Skowronski, X.R. Huang, M. Dudley, CVD Growth of 3C-SiC on 4H/6H Mesas, Chem. Vap. Depos. 12 (2006) 531–540. https://doi.org/10.1002/cvde.200506460

[59] V. Jokubavicius, G.R. Yazdi, R. Liljedahl, I.G. Ivanov, J. Sun, X. Liu, P. Schuh, M. Wilhelm, P. Wellmann, R. Yakimova, M. Syväjärvi, Single Domain 3C-SiC Growth on Off-Oriented 4H-SiC Substrates, Cryst. Growth Des. 15 (2015) 2940–2947. https://doi.org/10.1021/acs.cgd.5b00368

[60] P. Wu, M. Yoganathan, I. Zwieback, Y. Chen, M. Dudley, Characterization of Dislocations and Micropipes in 4H n^+ SiC Substrates, in: Silicon Carbide Relat. Mater. 2007, Trans Tech Publications, 2009: pp. 333–336.

[61] B. Kallinger, S. Polster, P. Berwian, J. Friedrich, G. Müller, A.N. Danilewsky, A. Wehrhahn, A.-D. Weber, Threading dislocations in n-and p-type 4H-SiC material analyzed by etching and synchrotron X-ray topography, J. Cryst. Growth. 314 (2011) 21–29. https://doi.org/10.1016/j.jcrysgro.2010.10.145

[62] M. Na, I.H. Kang, J.H. Moon, W. Bahng, Role of the oxidizing agent in the etching of 4H-SiC substrates with molten KOH, J. Korean Phys. Soc. 69 (2016) 1677–1682. https://doi.org/10.3938/jkps.69.1677

[63] P.H.L. Notten, J.E.A.M. Meerakker, J.J. Kelly, Etching of III-V semiconductors: an electrochemical approach, Elsevier Science Ltd, 1991.

[64] J.S. Shor, R.M. Osgood, Broad-Area Photoelectrochemical Etching of n-Type Beta-SiC, J. Electrochem. Soc. 140 (1993) L123-L125. https://doi.org/10.1149/1.2220722

[65] Y. Ke, R.P. Devaty, W.J. Choyke, Comparative columnar porous etching studies on n-type 6H SiC crystalline faces, Phys. Status Solidi. 245 (2008) 1396–1403. https://doi.org/10.1002/pssb.200844024

[66] A.O. Konstantinov, C.I. Harris, E. Janzen, Electrical properties and formation mechanism of porous silicon carbide, Appl. Phys. Lett. 65 (1994) 2699–2701. https://doi.org/10.1063/1.112610

[67] Y. Shishkin, W.J. Choyke, R.P. Devaty, Photoelectrochemical etching of n-type 4H silicon carbide, J. Appl. Phys. 96 (2004) 2311–2322. https://doi.org/10.1063/1.1768612

[68] G. Gautier, F. Cayrel, M. Capelle, J. Billoué, X. Song, J.-F. Michaud, Room light anodic etching of highly doped n-type 4 H-SiC in high-concentration HF electrolytes: Difference between C and Si crystalline faces, Nanoscale Res. Lett. 7 (2012) 367. https://doi.org/10.1186/1556-276X-7-367

[69] Y. Ke, R.P. Devaty, W.J. Choyke, Self-ordered nanocolumnar pore formation in the photoelectrochemical etching of 6H SiC, Electrochem. Solid-State Lett. 10 (2007) K24-K27. https://doi.org/10.1149/1.2735820

[70] P. Newby, J.-M. Bluet, V. Aimez, L.G. Fréchette, V. Lysenko, Structural properties of porous 6H silicon carbide, Phys. Status Solidi. 8 (2011) 1950–1953. https://doi.org/10.1002/pssc.201000222

[71] G. Gautier, J. Biscarrat, D. Valente, T. Defforge, A. Gary, F. Cayrel, Systematic Study of Anodic Etching of Highly Doped N-type 4H-SiC in Various HF Based Electrolytes, J. Electrochem. Soc. 160 (2013) D372-D379. https://doi.org/10.1149/2.082309jes

[72] S. Soloviev, T. Das, S.T. S., Structural and Electrical Characterization of Porous Silicon Carbide Formed in n-6H-SiC Substrates, Electrochem. Solid-State Lett. 6 (2003) G22–G24. https://doi.org/10.1149/1.1534733

[73] W. Lu, Y. Ou, P.M. Petersen, H. Ou, Fabrication and surface passivation of porous 6H-SiC by atomic layer deposited films, Opt. Mater. Express. 6 (2016) 1956–1963. https://doi.org/10.1364/OME.6.001956

[74] M. Leitgeb, C. Zellner, M. Schneider, U. Schmid, A Combination of Metal Assisted Photochemical and Photoelectrochemical Etching for Tailored Porosification of 4H SiC Substrates, ECS J. Solid State Sci. Technol. 5 (2016) P556--P564. https://doi.org/10.1149/2.0041610jss

[75] M. Leitgeb, C. Zellner, C. Hufnagl, M. Schneider, S. Schwab, H. Hutter, U. Schmid, Stacked Layers of Different Porosity in 4H SiC Substrates Applying a Photoelectrochemical Approach, J. Electrochem. Soc. 164 (2017) E337–E347. https://doi.org/10.1149/2.1081712jes

[76] M. Leitgeb, A. Backes, C. Zellner, M. Schneider, U. Schmid, Communication-The Role of the Metal-Semiconductor Junction in Pt-Assisted Photochemical Etching of Silicon Carbide, ECS J. Solid State Sci. Technol. 5 (2016) P148–P150. https://doi.org/10.1149/2.0021603jss

[77] J. Hassan, J.P. Bergman, A. Henry, E. Janzén, In-situ surface preparation of nominally on-axis 4H-SiC substrates, J. Cryst. Growth. 310 (2008) 4430–4437. https://doi.org/10.1016/j.jcrysgro.2008.06.083

[78]   C. Hallin, F. Owman, P. Mårtensson, A. Ellison, A. Konstantinov, O. Kordina, E. Janzén, In situ substrate preparation for high-quality SiC chemical vapour deposition, J. Cryst. Growth. 181 (1997) 241–253. https://doi.org/10.1016/S0022-0248(97)00247-9

[79]   H. Lee, B. Park, S. Jeong, S. Joo, H. Jeong, The effect of mixed abrasive slurry on CMP of 6H-SiC substrates, J. Ceram. Process. Res. 10 (2009) 378.

[80]   L. Zhou, V. Audurier, P. Pirouz, J.A. Powell, Chemomechanical Polishing of Silicon Carbide, J. Electrochem. Soc. 144 (1997) L161–L163. https://doi.org/10.1149/1.1837711

[81]   H. Deng, K. Endo, K. Yamamura, Competition between surface modification and abrasive polishing: a method of controlling the surface atomic structure of 4H-SiC (0001), Sci. Rep. 5 (2015) 8947. https://doi.org/10.1038/srep08947

[82]   M. Anikin, R. Madar, Temperature gradient controlled SiC crystal growth, Mater. Sci. Eng. B. 46 (1997) 278–286. https://doi.org/10.1016/S0921-5107(96)01993-9

[83]   S.P. Lebedev, V.N. Petrov, I.S. Kotousova, A.A. Lavrentev, P.A. Dementev, A.A. Lebedev, N. Titkov, Formation of Periodic Steps on 6H-SiC (0001) Surface by Annealing in a High Vacuum, Mater. Sci. Forum. 679 (2011) 437. https://doi.org/10.4028/www.scientific.net/MSF.679-680.437

[84]   T. Nishiguchi, S. Ohshima, S. Nishino, Thermal Etching of 6H–SiC Substrate Surface, Jpn. J. Appl. Phys. 42 (2003) 1533. https://doi.org/10.1143/JJAP.42.1533

[85]   N.G. van der Berg, J.B. Malherbe, A.J. Botha, E. Friedland, Thermal etching of SiC, Appl. Surf. Sci. 258 (2012) 5561–5566. https://doi.org/10.1016/j.apsusc.2011.12.132

[86]   I. Swiderski, Thermal etching of α-SiC crystals in argon, J. Cryst. Growth. 16 (1972) 1–9. https://doi.org/10.1016/0022-0248(72)90079-6

[87]   V. Jokubavicius, G.R. Yazdi, I.G. Ivanov, Y. Niu, A. Zakharov, T. Iakimov, M. Syväjärvi, R. Yakimova, Surface engineering of SiC via sublimation etching, Appl. Surf. Sci. 390 (2016). https://doi.org/10.1016/j.apsusc.2016.08.149

[88]   G. Honstein, C. Chatillon, F. Baillet, Thermodynamic approach to the vaporization and growth phenomena of SiC ceramics. I. SiC and SiC–SiO_2 mixtures under neutral conditions, J. Eur. Ceram. Soc. 32 (2012) 1117–1135. https://doi.org/10.1016/j.jeurceramsoc.2011.11.032

[89]   Y.A. Vodakov, A.D. Roenkov, M.G. Ramm, E.N. Mokhov, Y.N. Makarov, Use of Ta-Container for Sublimation Growth and Doping of SiC Bulk Crystals and

Materials Research Forum LLC
doi: http://dx.doi.org/10.21741/9781945291852

Epitaxial Layers, Phys. Status Solidi. 202 (1997) 177–200.
https://doi.org/10.1002/1521-3951(199707)202:1<177::AID-PSSB177>3.0.CO;2-I

[90] F.R. Chien, S.R. Nutt, W.S. Yoo, T. Kimoto, H. Matsunami, Terrace growth and polytype development in epitaxial β-SiC films on α-SiC (6H and 15R) substrates, J. Mater. Res. 9 (1994) 940–954. https://doi.org/10.1557/JMR.1994.0940

[91] V. Heine, C. Cheng, R.J. Needs, The Preference of Silicon Carbide for Growth in the Metastable Cubic Form, J. Am. Ceram. Soc. 74 (1991) 2630–2633. https://doi.org/10.1111/j.1151-2916.1991.tb06811.x

[92] T. Matsumoto, J. Takahashi, T. Tamaki, T. Futagi, H. Mimura, Y. Kanemitsu, Blue-green luminescence from porous silicon carbide, Appl. Phys. Lett. 64 (1994) 226–228. https://doi.org/10.1063/1.111979

[93] T.L. Rittenhouse, P.W. Bohn, T.K. Hossain, I. Adesida, J. Lindesay, A. Marcus, Surface-state origin for the blueshifted emission in anodically etched porous silicon carbide, J. Appl. Phys. 95 (2004) 490–496. https://doi.org/10.1063/1.1634369

[94] S. Kamiyama, T. Maeda, Y. Nakamura, M. Iwaya, H. Amano, I. Akasaki, H. Kinoshita, T. Furusho, M. Yoshimoto, T. Kimoto, J. Suda, A. Henry, I.G. Ivanov, J.P. Bergman, B. Monemar, T. Onuma, S.F. Chichibu, Extremely high quantum efficiency of donor-acceptor-pair emission in N-and-B-doped 6H-SiC, J. Appl. Phys. 99 (2006) 93108. https://doi.org/10.1063/1.2195883

[95] S. Kamiyama, M. Iwaya, T. Takeuchi, I. Akasaki, M. Syväjärvi, R. Yakimova, Fluorescent SiC and its application to white light-emitting diodes, J. Semicond. 32 (2011) 13004. https://doi.org/10.1088/1674-4926/32/1/013004

[96] T. Nishimura, K. Miyoshi, F. Teramae, M. Iwaya, S. Kamiyama, H. Amano, I. Akasaki, High efficiency violet to blue light emission in porous SiC produced by anodic method, Phys. Status Solidi. 7 (2010) 2459–2462. https://doi.org/10.1002/pssc.200983908

[97] W. Lu, Y. Ou, E.M. Fiordaliso, Y. Iwasa, V. Jokubavicius, M. Syväjärvi, S. Kamiyama, P.M. Petersen, H. Ou, White Light Emission from Fluorescent SiC with Porous Surface, Sci. Rep. 7 (2017) 9798. https://doi.org/10.1038/s41598-017-10771-7

[98] M. Leitgeb, C. Zellner, M. Schneider, U. Schmid, Porous single crystalline 4H silicon carbide rugate mirrors, APL Mater. 5 (2017) 106106. https://doi.org/10.1063/1.5001876

[99]　A.J. Rosenbloom, S. Nie, Y. Ke, R.P. Devaty, W.J. Choyke, Columnar morphology of porous silicon carbide as a protein-permeable membrane for biosensors and other applications, in: Mater. Sci. Forum, 2006: pp. 751–754. https://doi.org/10.4028/0-87849-425-1.751

[100]　A.J. Rosenbloom, D.M. Sipe, Y. Shishkin, Y. Ke, R.P. Devaty, W.J. Choyke, Nanoporous SiC: A candidate semi-permeable material for biomedical applications, Biomed. Microdevices. 6 (2004) 261–267. https://doi.org/10.1023/B:BMMD.0000048558.91401.1d

[101]　Y.-W. Son, M.L. Cohen, S.G. Louie, Half-metallic graphene nanoribbons, Nature. 444 (2006) 347. https://doi.org/10.1038/nature05180

[102]　V. Barone, O. Hod, G.E. Scuseria, Electronic structure and stability of semiconducting graphene nanoribbons, Nano Lett. 6 (2006) 2748–2754. https://doi.org/10.1021/nl0617033

[103]　J. Baringhaus, M. Ruan, F. Edler, A. Tejeda, M. Sicot, A. Taleb-Ibrahimi, A.-P. Li, Z. Jiang, E.H. Conrad, C. Berger, C. Tegenkamp, W.A. de Heer, Exceptional ballistic transport in epitaxial graphene nanoribbons, Nature. 506 (2014) 349. https://doi.org/10.1038/nature12952

[104]　A. Stöhr, J. Baringhaus, J. Aprojanz, S. Link, C. Tegenkamp, Y. Niu, A.A. Zakharov, C. Chen, J. Avila, M.C. Asensio, others, Graphene Ribbon Growth on Structured Silicon Carbide, Ann. Phys. 529 (2017) 1700052. https://doi.org/10.1002/andp.201700052

[105]　M.S. Nevius, F. Wang, C. Mathieu, N. Barrett, A. Sala, T.O. Mentes, A. Locatelli, E.H. Conrad, The bottom-up growth of edge specific graphene nanoribbons, Nano Lett. 14 (2014) 6080–6086. https://doi.org/10.1021/nl502942z

[106]　M. Sprinkle, M. Ruan, Y. Hu, J. Hankinson, M. Rubio-Roy, B. Zhang, X. Wu, C. Berger, W.A. De Heer, Scalable templated growth of graphene nanoribbons on SiC, Nat. Nanotechnol. 5 (2010) 727. https://doi.org/10.1038/nnano.2010.192

Advancing Silicon Carbide Electronics Technology I   Materials Research Forum LLC
Materials Research Foundations **37** (2018)   doi: http://dx.doi.org/10.21741/9781945291852

CHAPTER 2

Processing and Characterisation of Ohmic Contacts to Silicon Carbide

K. Vasilevskiy[1]*, K. Zekentes[2], N. Wright[1]

[1] Newcastle University, Newcastle upon Tyne, United Kingdom

[2] Microelectronics Research Group (MRG), Institute of Electronic Structure and Laser (IESL), Foundation for Research & Technology Hellas (FORTH), Heraklion, Greece

* konstantin.vasilevskiy@newcastle.ac.uk

Abstract

This chapter reports on formation and characterization of ohmic contacts to silicon carbide. At first, the theory of ohmic contacts is briefly described with special attention to dependence of contact resistance on semiconductor parameters and to the differences between ohmic contacts to silicon carbide and silicon. Then, different contact resistivity measurement techniques are discussed with primary emphasis upon the transfer length method (TLM). TLM limitations, accuracy and optimized test structure design are considered in detail. Recent progress in the development of ohmic contacts is reviewed with more detailed description of commonly used nickel and aluminum-titanium contacts to *n*- and *p*-type silicon carbide, respectively. Protection, overlaying and thermal stability of ohmic contacts to SiC are discussed as well as compatibility of ohmic contacts formation with SiC device processing. Finally, the requirements to further improvements in ohmic contacts fabrication and characterization are outlined.

Keywords

Silicon Carbide, Ohmic Contact, Transition Line Model, Transfer Length Method, Contact Resistivity, Metal Silicides, Metal Carbides, Diffusion Barrier, Schottky Barrier, High Temperature Electronics

Contents

List of used symbols and abbreviations

A fitting parameter in TLM;

A^* ($A \cdot cm^{-2} \cdot K^{-2}$) effective Richardson constant;

A_0 ($A \cdot cm^{-2} \cdot K^{-2}$) Richardson constant for free electrons;

B fitting parameter in TLM;

c (J/g/K) specific heat capacity;

c_2 Fermi level pinning parameter;

c_3 Fermi level pinning parameter;

d_i (μm) separation distance between contacts *i* and *i-1* in a TLM test structure;

d_i (μm) contact separation distance;

d_{max} (μm) maximum contact separation distance;

E_{00} (eV) tunneling parameter;

E_C (eV) minimum conduction band energy;

E_G (eV) band gap energy;

E_N (eV) surface charge neutrality level;

E_P (J/cm^2) laser pulse energy density;

E_V (eV) maximum valence band energy;

F (V/cm) electric field strength;

F_{max} (V/cm) maximum electric field strength in a Schottky barrier;

$\hbar$ (eV·s) reduced Plank's constant, 6.582×10^{-16};

I (A) current;

J (A/cm^2) current density;

J_S (A/cm^2) saturation current density;

k_B (J/K) Boltzmann's constant, 1.38066×10^{-23};

L_y (μm) intersection of TLM fitted line with d axis;

L (μm) contact pad length in a TLM test structure;

L_T (μm) transfer length;

$l_{x,y,z}$ direction cosines of the carriers flux relative to the principle axes of the constant energy ellipsoid;

M number of contact pad spacings in a TLM test structure;

M_C number of equivalent valleys in conduction band;

m_0 (kg) electron rest mass, 9.109×10^{-31};

m_e (kg) effective mass of electrons;

m_{ei} (kg) effective mass of electrons in a single valley;

$m_{e\perp}$ (kg) transverse effective mass of electrons (perpendicular to the <0001> direction in SiC);

$m_{e\parallel}$ (kg) longitudinal effective mass of electrons (parallel to the <0001> direction in SiC);

m_{tun} (kg) tunneling effective mass;

m_{hh} (kg) effective mass of heavy holes;

m_{lh} (kg) effective mass of light holes;

$m_{h\parallel}$ (kg) longitudinal effective mass of holes (parallel to the <0001> direction in SiC);

$m_{h\perp}$ (kg) transverse effective mass of holes (perpendicular to the <0001> direction in SiC);

m_h (kg) effective hole mass;

$m_{e\,M\Gamma}$(kg) effective electron mass in a single valley along the corresponding direction;

$m_{e\,MK}$ (kg) effective electron mass in a single valley along the corresponding direction;

$m_{e\,ML}$(kg) effective electron mass in a single valley along the corresponding direction;

$m_{x,y,z}$ (kg) components of the effective mass tensor;

n idealityfactor;

N_A (cm^{-3}) acceptor concentration;

N_C (cm^{-3}) effective density of states in conduction band at RT;

N_D (cm^{-3}) donor atomic concentration;

N_V (cm^{-3}) effective density of states in valence band at RT;

q (C) elementary charge, 1.6022×10^{-19};

r (cm) contact radius;

R (Ω) measured resistance;

R_d (Ω) differential resistance;

R_{DS} (Ω) drain-source resistance in FET;

R_{Cf} (Ω) front contact resistance;

R_{Ce} (Ω) end contact resistance;

R_{max} (Ω) maximum resistance measured in a specific TLM test structure;

R_{Sh} (Ω/sq.) semiconductor layer sheet resistance;

R_{Sk} (Ω/sq.) sheet resistance of semiconductor layer under contact;

S (cm^2) area;

T (°C, K) temperature;

t (s) time;

t_P (s) laser pulse duration;

V (V) absolute value of voltage bias;

V_{bn} (V) diffusion potential or built-in voltage of a Schottky barrier;

W (μm) width of TLM test structure termination;

x_S (cm) semiconductor layer thickness;

x_T (cm) heat penetration depth;

Z (μm) width of contact pads in a TLM test structure;

Z_{opt} (μm) optimized width of contact pads in a TLM test structure;

$\Delta\phi$ (eV) Schottky barrier image-force lowering due to electric field in the interfacial layer;

$\Delta\phi_C$ (eV) Schottky barrier image-force lowering;

ΔX general error expressions for the parameter X that can represent either random or systematic contributions;

δX systematic error of the parameter X, which is a consistent shift of the mean X value that cannot be reduced by taking larger numbers of data points;

ε relative dielectric constant;

ε_0 (F/cm) vacuum permittivity, 8.854×10^{-14};

θ (arcsec) incident beam angle in XRD;

κ (g/cm^3) material density;

λ (W/cm/K) heat conductance;

ρ_B (Ω·cm) semiconductor bulk resistivity;

ρ_C (Ω·cm^2) specific contact resistance (or contact resistivity);

σX standard deviation used to quantify a random error of parameter X;

ϕ_B (eV) Schottky barrier height;

ϕ_M (eV) metal work function;

ϕ_S (eV) semiconductor work function;

ϕ_{SS} (eV) E_N-E_V;

χ_S (eV) semiconductor electron affinity;

χ_T (cm^2/s) temperature conductance, $\lambda/c\kappa$;

AES	Auger electron spectroscopy;
CBKR	cross bridge Kelvin resistance;
CSM	Cox and Strack method;
CTLM	circular modification of TLM;
CVD	chemical vapor deposition;
EDM	electrical discharge machining;
FE	field emission;
IC	integrated circuit;
ICP	inductively coupled plasma;
JFET	junction field-effect transistors;
LTLPE	low temperature liquid phase epitaxy;
MESFET	metal-semiconductor field-effect transistors;
MOSFET	metal-oxide-semiconductor field-effect transistors;
PECVD	plasma enhanced chemical vapour deposition;
PDA	post deposition annealing;
PIA	post implantation annealing;
PLA	pulse laser annealing;
PLD	pulsed laser deposition;
PSU	power supply unit;
RIE	reactive-ion etching;
rms	root-mean square;
RT	room temperature;
RTP	rapid thermal processing;
SBC	Schottky barrier contacts;

SCR   space charge region;

SIMS   secondary ion mass spectrometry;

SMU   source-measure unit;

TE   thermionic emission;

TEC   thermal expansion coefficient;

TFE   thermionic-field emission;

TEM   transmission electron microscopy;

TI-VJFET   trenched and implanted vertical-channel junction field-effect transistors;

TLM   transfer length method;

UHV   ultra-high vacuum;

XPS   x-ray photoelectron spectroscopy;

XRD   x-ray diffraction.

1. Introduction

Fabrication of ohmic contacts to silicon carbide was addressed by R. Hall for the first time in 1958 [1]. Ohmic contacts to p-type SiC Lely crystals [2] were fabricated by fusing 200 µm thick silicon-aluminum (1:1) alloy at temperature about 1700 °C. Alloys of silicon with approximately one percent phosphorous have been used to fabricate non-rectifying junctions on n-type SiC. Silicon carbide was dissolved to a depth of a few microns during that procedure. Contacts were not characterized quantitatively but only their non-rectifying behavior was addressed. I-V characteristics of fabricated rectifiers were measured and fitted by a theoretical I-V expression for ideal p-n junction using series resistance as a fitting parameter. It was found to be of 4.5 and 1.5 Ω at temperature of 27 and 500 °C, respectively, for device area of 3×10^{-3} cm^2. This results in the upper bound for contacts resistivity of 13.5×10^{-3} and 4.5×10^{-3} $\Omega \cdot$cm^2 at 27 and 500 °C, respectively. For a long period of time the ohmic contact resistivity of this scale as well as their harsh fabrication conditions remained acceptable since the main application of SiC was considered to be a material for light emitting diodes and device structures were fabricated on Lely crystals grown by sublimation at much higher temperatures.

Invention of the modified Lely method for seeded growth of SiC crystals in the late 1970s [3] and development of the epitaxial growth of SiC layers in the next decade [4] resulted in availability of thin SiC layers with controlled polytype, thickness and doping level for fabrication of SiC devices. This progress in material growth necessitated development of more advanced techniques for fabrication and characterization of ohmic contacts to silicon carbide with accurate quantitative characteristics and less damage to the underlying thin SiC layer. The first review of ohmic contacts to SiC was published by Harris *et al.* in 1993[5]. Then, ohmic contacts to *n*-and *p*-type 6*H*-SiC were critically reviewed by L.M. Porter and R.F. Davis [6] in 1995 and by J. Crofton *et al.* in 1997 [7].

During the time preceding these reviews, 6*H*-SiC was the main polytype grown by Lely and modified Lely methods. 6*H*-SiC epitaxial layers grown by different methods were also available and their main use was considered to be in a fabrication of blue light emitting diodes. To the end of 1990s, III-N LEDs were developed and SiC was replaced in this application niche. Further progress in SiC growth and device fabrication technology was driven by the rising demand for efficient high power semiconductor devices. For this application, the 4*H* polytype is a better choice than 6*H* due to higher electron and hole mobilities. Furthermore, 4*H*-SiC has higher avalanche breakdown field and electron mobility is almost isotropic in it. Epitaxial 4*H*-SiC layers became commercially available and since then the development of SiC post-growth technology was mainly focused in processing of 4*H*-SiC devices.

In 2005, F. Roccaforte *et al.* [8] published the first review of ohmic contacts to 4*H*-SiC as well as to 6*H*-SiC. More recently, Z. Wang *et al.* (2016) [9] published a comprehensive review of ohmic contacts to both SiC polytypes with the majority of data provided for 4*H*-SiC. These reviews still remain topical and representative up to now, therefore this book chapter aims not to duplicate them but rather to provide a practical guide for understanding, fabrication and characterization of ohmic contacts to SiC mainly focusing on 4*H* polytype. The chapter starts with a brief description of metal-semiconductor contact theory and methods of contact resistivity measurement with particular emphasis on their application to SiC. This will be followed by practical advice on the fabrication of ohmic contacts to SiC with more detailed description of commonly used nickel and aluminum-titanium based contacts to *n*- and *p*-type silicon carbide, respectively. Although this chapter does not pretend to provide a full and comprehensive review of all published data on ohmic contacts to SiC, it includes critical analysis of data provided by aforementioned reviews as well as those more recently published.

2.   Ohmic contacts: definition, theory and dependence on semiconductor parameters

Metal-semiconductor contacts are mandatory and significant components of all semiconductor devices. They can be either rectifying or ohmic. An ohmic contact is defined as a contact for which current-voltage (I-V) characteristics are determined by the resistivity of the semiconductor specimen or by the behavior of the device of which the contact forms part, rather than by the characteristics of the contact itself [10]. Ohmic contacts must be able to supply the necessary device current, and the voltage drop across the contact should be small compared to the voltage drops across the active device regions. They should not inject minority carriers into the semiconductor and additionally, should be stable electrically and mechanically.

The main characteristics of an ohmic contact is its resistivity (or specific resistance) defined as [11]:

$$\rho_C = \left.\frac{\partial V}{\partial J}\right|_{V=0} \tag{1a}$$

where J is the current density. It is assumed that the contact layer, metal-semiconductor interface and semiconductor layer under the contact are uniform and contact resistivity does not depend on contact area (S). To reflect these conditions, Eq. 6.1a is sometimes replaced with a more strict definition:

$$\rho_C = \lim_{S\to 0} \left(\left.\frac{\partial V}{\partial J}\right|_{V=0}\right) \tag{1b}$$

The definition of ρ_C at zero voltage bias came from a theoretical consideration of ohmic contacts as a particular case of Schottky barrier contacts (SBC) which will be defined later in this section. Indeed, electrical current through the metal-semiconductor contact with Schottky barrier is determined by different physical processes depending on material parameters and doping level of the semiconductor. For all these processes, theoretical expressions for $J(V)$ dependence are complicated and the derivative $\partial V/\partial J$ can be expressed in closed form only for a marginal case of $V \to 0$. That is why the ρ_C parameter defined by Eq. 1a was chosen as a figure of merit for theoretical comparison of metal-semiconductor contacts with different Schottky barrier heights.

In practice, all semiconductor devices operate at some current or voltage bias and have a non-zero current density passing through contacts and, consequently, non-zero voltage drop on them. Therefore, the dependence of contact resistivity on current density or voltage should be specified in ohmic contacts to be usable in different devices operating at different current densities. The simplest case of this dependence is the absence of any dependence. That is why it is commonly accepted that the first sign of an ohmic contact is an "ohmic behavior" which means that the contact demonstrates non-rectifying linear (or "quasi-linear") $I–V$ characteristic and ρ_C value does not depend on V although it is not required by the classical definition of ohmic contact and by ρ_C definition given by Eq. 1.

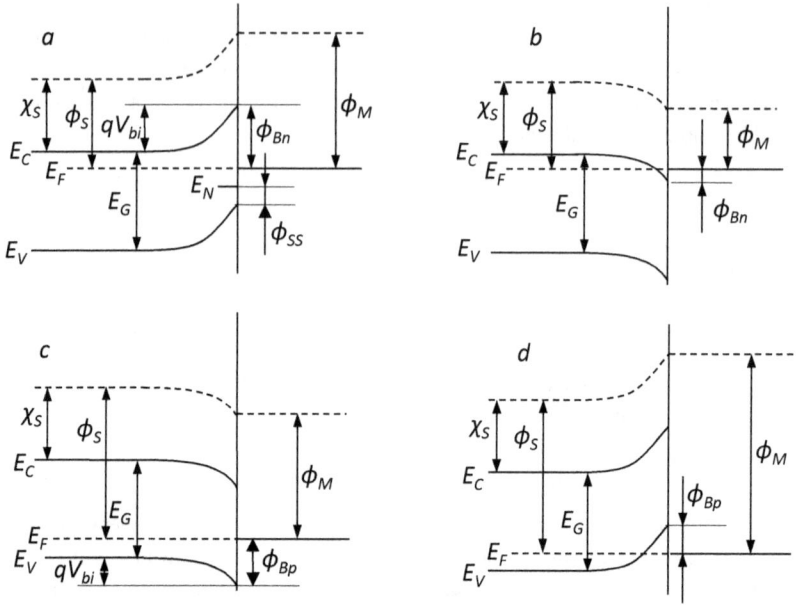

Figure 1. *Energy band diagram for metal-semiconductor interface (a, b) for n-type semiconductor and (c, d) for p-type semiconductor.*

It is a common statement that it is difficult to fabricate ohmic contacts to silicon carbide because it is a wide band gap semiconductor. To find if it is correct, let's consider Fig. 1 which shows the energy band diagrams for perfect metal-semiconductor contacts (when materials at both sides of a contact retain their bulk properties right up to the metallurgical interface). The Fermi levels at both sides of the contact have to be coincident to each other in the case of thermal equilibrium. As far as a metal and a semiconductor have different work functions, the Fermi levels in the semiconductor is either above or below the metal Fermi level before the materials are brought in intimate contact. Electrons have to move to the material with lower Fermi level until the Fermi level becomes a constant through the system. This creates either depleted of enhanced space charge region (SCR) in the semiconductor (the SCR in the metal is negligibly thin due to very high electron concentration). The electric field in SCR prevents further movement of carriers. It is rising linear and creates the parabolic potential barrier, V_{bi}, at the semiconductor side of the contact (supposing the uniform charge distribution in

SCR). The contacts with this barrier are called as a Schottky barrier contacts. The Schottky barrier height is measured from the Fermi level and is defined as:

$$\phi_{Bn} = \phi_M - \chi_S \tag{2a}$$

$$\phi_{Bp} = \chi_S + E_G - \phi_M = E_G - \phi_{Bn} \tag{2b}$$

for n- and p-type semiconductor, respectively. Note that it is supposed that the electron affinity (χ_S) in semiconductors does not depend on the type of conductivity. In the case of $\phi_M < \chi_S$ and $\phi_M > \chi_S + E_G$ for n- and p- type semiconductor, respectively, the Schottky barrier heights are negative and the contacts become ohmic due to the energy band alignment as it is shown in Fig. 1b and Fig. 1c for n- and p-type semiconductor, respectively. Indeed, there are no barriers for carriers moving from the semiconductor to the metal. When the opposite bias is applied, the inversed layer of semiconductor services as a carrier reservoir for current in reverse direction. Therefore, the current in both directions is entirely defined by bulk semiconductor properties as it is required by the definition of ohmic contact.

Fig. 2 shows Schottky barrier heights depending on metal work function calculated by Eq. 2 for metal / n-type 4H-SiC (solid line 1 in Fig. 2a) and for metal / p-type 4H-SiC (solid line 1 in Fig. 2b). Table 1 lists selected parameters of silicon and silicon carbide used for these calculations. Important parameters of selected metals and metal compounds which are used or are potentially useable for fabrication of ohmic contacts, contacts capping and protecting layers are listed in Table 2.

4H-SiC has relatively low value of χ_S (3.2 eV), only rare earth and alkali metals have lower work functions which are required to form contacts to n-type 4H-SiC with negative ϕ_{Bn} by the energy band alignment according to Eq. 2a. These metals are not practically usable due to their chemical activity. In the case of contacts to p-type 4H-SiC, the wide band gap of SiC overcompensates its low electron affinity making it impossible to form contacts to p-type 4H-SiC with negative ϕ_{Bp} by the energy band alignment since there is no materials with work function higher than 6.43 eV required by Eq. 2b. Fig. 2 also shows Schottky barriers for n- and p-type Si calculated by Eq. 2 for comparison.

Advancing Silicon Carbide Electronics Technology I   Materials Research Forum LLC
Materials Research Foundations **37** (2018)   doi: http://dx.doi.org/10.21741/9781945291852

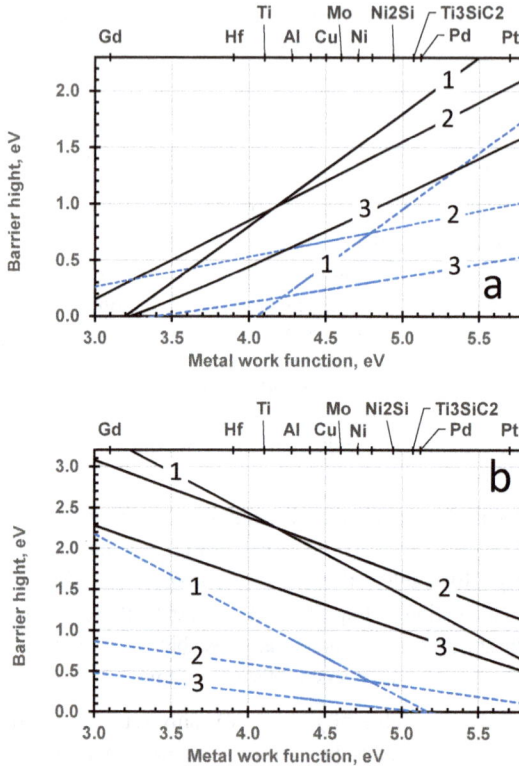

Figure 2.   Schottky barrier height depending on metal work function in (a) metal – n-type and (b) metal – p-type semiconductor contacts. Solid lines - 4H-SiC; dashed lines – Si;

lines marked "1": ϕ_{Bn} and ϕ_{Bp} as calculated by Eq. 6.2;

lines marked "2": ϕ_{Bn} and ϕ_{Bp} calculated with taking into account Fermi level pinning for lightly doped semiconductors; and

lines marked "3": ϕ_{Bn} and ϕ_{Bp} calculated with taking into account Fermi level pinning and barrier image-force lowering for semiconductors doped up to the solubility limit.

Table 1. Selected material parameters of silicon and silicon carbide used for calculations of barrier heights and contact resistivity.

	Si	4H-SiC	6H-SiC
E_g at RT (eV)	1.12 [12]	3.23 [13]	3.0 [13]
χ_S (eV)	4.05 [12]	3.2 [14]	3.8 [15]
ε	11.9 [13]	9.7 [13]	9.7 [13]
N_C at RT (cm^{-3})	2.8×10^{19} [12]	1.6×10^{19} [13]	8.7×10^{19} [13]
N_V at RT (cm^{-3})	1.04×10^{19} [12]	2.4×10^{19} [13]	2.4×10^{19} [13]
M_C	6 [16]	3 [13]	6 [17]
$m_{e\,M\Gamma}/m_0$		0.58 [18]	0.75 [17]
$m_{e\,MK}/m_0$		0.31[18]	0.24 [17]
$m_{e\,ML}/m_0$		0.33[18]	1.83 [17]
$m_{e\perp}/m_0$	0.19 [16]	0.42 [19]	0.42 [13]
$m_{e\parallel}/m_0$	0.98 [16]	0.29 [19]	2 [13]
m_e/m_0	2.05 [20]	1.27	2.55
m_{hh}/m_0	0.49 [12]		
m_{hl}/m_0	0.16 [12]		
$m_{h\parallel}/m_0$		1.75 [21]	1.85 [21]
$m_{h\perp}/m_0$		0.66 [21]	0.66 [21]
m_h/m_0	0.65 [20]	1.07	1.10
c_2	0.27 [12]	0.70 [22]	0.47-0.63 [22]
c_3	-0.55 [12]	-1.95 [22]	
ϕ_{SS}, (eV)	0.3 [12]	2.26 [22]	
Max. donor impurity solubility (cm^{-3})	1.2×10^{21} (P) [23]	5×10^{20} (N by modified Lely method) [24] 2.8×10^{18} (P by CVD) [21] $\sim 10^{20}$ (P by ion implantation) [25]	
Max. acceptor impurity solubility (cm^{-3})	5.7×10^{20} (B) [26]	7.0×10^{20} (Al by LTLPE) [27] 2.0×10^{21} (Al by sublimation sandwich method) [28]	

Restrictions on metals work functions required to have negative barrier heights in contacts to silicon are not as tough as for silicon carbide since silicon has higher electron affinity and lower band gap. According to this model, metals like titanium and platinum have to form ohmic contacts to *n*- and *p*-type silicon, respectively, independently on its doping level.

Table 2. Selected parameters of some metals and metal compounds which are used or potentially usable for fabrication ohmic contacts, contacts capping and protecting.

Metal	Electrical resistivity ($\mu\Omega\cdot$cm)	TEC, $\times 10^6$ (K^{-1})	Work function or electron affinity (eV)	Melting point (°C)
Hf	33.1	5.9	3.9	2233
Ti	54	8.9	4.1	1660
Si	2.5	4.7-7.67	4.15	1410
Al	2.67	23.5	4.2	660.4
Mo	5.7	5.1	4.2	2625
Cr	13.2	6.5	4.4	1890
Cu	1.69	17.0	4.5	1083
W	5.4	4.5	4.55	3410
Ni	6.9 [29]	13.3 [29]	4.71 [30]	1453
Au	2.2	14.1	4.8	1063
C	1375	0.5	4.8	3650
Co	6.34	12.5	5.0	1495
Pt	10.6	8.8	5.7	1768
Ni_2Si	24-30 [31]	12 [31]	4.94 [32]	1255 [31]
NiSi	10.5-18 [31]		4.82 [32, 33]	992 [31]
$NiSi_2$	34-50 [31]		5.03 [32]	990 [31]
TaC	36 [34]	8.4 [35]	4.22 [36]	3880
$TaSi_2$	36.7 [37]	14 [38]	4.71 [39]	2200
TiC	126 [40]	4.1-7.7	3.7 (110); 3.8 (100); 4.7 (111) [41]	3160
TiN	30-70	9.35	4.7 [42]	2930
Ti_3SiC_2	22 [43]	8 [43]	5.07 [44]	Stable up to 1400
WC	20	3.8 [45]	4.3 [46]	2870
4*H*- SiC (*n*-type)	20000	4.67	3.4	2830

The model above described is useful for understanding the physics of a metal-semiconductor contact and differences between Si and SiC but it is barely applicable in practice because it does not include surface states in the metal-semiconductor interface.

Advancing Silicon Carbide Electronics Technology I
Materials Research Foundations **37** (2018)

Materials Research Forum LLC
doi: http://dx.doi.org/10.21741/9781945291852

These surface states always exist even in a perfect "epitaxial" contact because of the breaking of the semiconductor crystal symmetry in the interface, dangling bonds or metal-semiconductor bonds. The model describing the energy band alignment in presence of surface states assumes that these states are separated from the metal by an interfacial layer [10] which is thin enough for electrons to readily tunnel through it but the drop of potential across this layer provides alignment of Fermi levels in metal and in surface states. The surface states may be either of donor type when the states above the Fermi level (not filled by electrons) are positively charged or of acceptor type when the states below Fermi level (filled by electrons) are negatively charged. Note that the type of a surface state does not depend of its position in the band gap and both donor and acceptor like surface states may be located either close to the top of valence band or to the bottom of the conductance band. The energy level at which the total charge of surface states is equal to zero is called a surface charge neutrality level (E_N). It is located at the distance of ϕ_{SS} from the top edge of the valence band as it is shown in Fig. 1a. In the presence of surface states, the surface charge has to be balanced by the additional charge in SCR and therefore the Fermi level position at the interface on the semiconductor side depends on the semiconductor doping, surface state density and surface charge neutrality level. The Schottky barrier height in this case can be expressed for an n-type semiconductor as [12]:

$$\phi_{Bn} = c_2(\phi_M - \chi_S) + (1 - c_2) \times (E_G - \phi_{SS}) - \Delta\phi \equiv c_2\phi_M + c_3 \tag{3a}$$

and for a p-type material [47]:

$$\phi_{Bp} = E_G - c_2(\phi_M - \chi_S) - (1 - c_2) \times (E_G - \phi_{SS}) - \Delta\phi \equiv E_G - \phi_{Bn} \tag{3b}$$

where c_2 and c_3 are the Fermi level pinning parameters and $\Delta\phi$ is a barrier lowering due to electric field in the interfacial layer. It is clear from Eqs. 3 that the Fermi level at the interface coincides with E_N ("pinned" to E_N) when $c_2=0$ and the Schottky barrier height does not depend on the metal work function in this case at all. The stronger the chemical bonding in a semiconductor crystal the lower is the effect of disordered bonds at the surface on its properties and hence Fermi level pinning should be weaker. In this case, c_2 tends to 1 and Eq. 3 reduces to Eq. 2. SiC is closer to ionic crystals with strong chemical bonding [48] while silicon is a covalent semiconductors with weaker chemical bonding and, hence, SiC has to have weaker Fermi level pinning in comparison to Si. Indeed, parameters c_2 and c_3 were determined experimentally and $c_{2(Si)}$ is remarkably lower than $c_{2(SiC)}$ as it is shown in Table 1.

The values of ϕ_{Bn} and ϕ_{Bp} calculated by Eq. 3 are shown in Fig. 2 by the lines marked with "2". It is evident that taking into account Fermi level pinning reduces the slope $\phi_{Bn}(\phi_M)$ and $\phi_{Bp}(\phi_M)$ dependences. This eliminates the possibility to get a contact with the negative Schottky barrier by means of energy band alignment both to silicon and to silicon carbide. Indeed, substituting c_2 and c_3 from Table 1 for 4H-SiC and assuming zero barriers heights results in contact materials work functions required for forming ohmic contacts by the energy band alignment $\phi_M < 2.8$ eV and $\phi_M > 7.4$ eV for n- and p-type 4H-SiC, respectively.

Eq. 3 includes Schottky barrier lowering due to the electric field in the interfacial layer ($\Delta\phi$). It increases with semiconductor doping and should lead to reduction of Schottky barriers in contacts to heavier doped semiconductors. The value of $\Delta\phi$ can be calculated if ϕ_{SS}, ϕ_B are known as well as density of surface states (N_{SS}), relative dielectric constant and thickness of the interfacial layer [10]. Unfortunately, the full set of these parameters is not available for SiC. The $\Delta\phi$ value was estimated not to exceed about 0.02 eV for moderately doped Si ($N_D \sim 10^{16}$cm^{-3}) and $\phi_{Bn}=0.5$ eV [10] which is an insignificant reduction of Schottky barrier height. It is also unlikely that it has a significant effect on the formation of ohmic contact to heavier doped SiC.

Another mechanism of Schottky barrier height reduction is image-force lowering. It is not shown in Fig. 1 but it can be calculated (for n-type semiconductor) as [10]:

$$\Delta\phi_C = q\sqrt{\frac{qF_{max}}{4\pi\varepsilon\varepsilon_0}} \tag{4}$$

where

$$F_{max} = \sqrt{\frac{2N_D}{\varepsilon\varepsilon_0}\left(\phi_{B0} - (E_C - E_F) - k_BT\right)} \tag{5}$$

is the maximum electric field strength in the Schottky barrier (at x=0) and ϕ_{B0} is the barrier height without lowering. In the non-degenerated n-type semiconductor, the Fermi level position ($E_C - E_F$) logarithmically depends on doping concentration (N_D) and the effective density of states in conduction band (N_C):

$$E_C - E_F = k_BT\ln\left(\frac{N_C}{N_D}\right) \tag{6}$$

Similar expression can be drawn for p-type semiconductor by replacing corresponding variables. This barrier image-force lowering at room temperature (RT) is shown in Fig. 3 for p- and n-type 4H-SiC as well as for n-type Si for comparison. The used ϕ_{B0} values were 0.92 eV for n-type 4H-SiC (Ti metallisation); 1.19 eV for p-type 4H-SiC (Pt metallisation) and 0.56 eV for n-type Si (Ti metallisation). This barrier lowering is shown at doping levels up to the corresponding impurity solubility limit and it is clear that it is very significant at high doping levels and is noticeably higher in SiC than in Si.

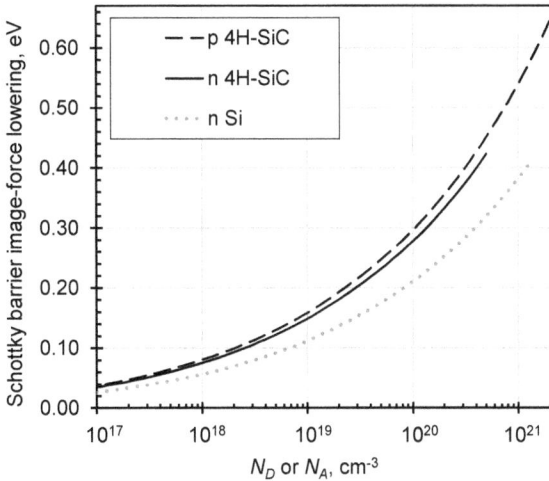

Figure 3.   Schottky barrier image-force lowering at room temperature (RT) depending on doping level for p and n-type 4H-SiC as well as for n-type Si for comparison.

The values of ϕ_{Bn} and ϕ_{Bp} calculated by Eq. 3 and reduced by $\Delta\phi_C$ are shown in Fig. 2 by the lines marked with "3". They were calculated for semiconductors doped up to the impurity solubility limit. Strictly speaking, this is not a correct procedure since Eq. 4 was derived for the metal-semiconductor contacts with zero interfacial states densities. However, this gives us an estimation of the maximum possible effect of image-force lowering on ohmic contacts formation. It is clearly seen from Fig. 2 that ϕ_{Bn} and ϕ_{Bp} for contacts to 4*H*-SiC doped up to the impurity solubility limits still remain positive over the full range of available metal work functions. Contact material with $\phi_M < 3.2$ eV and $\phi_M >$ 6.7 eV are required for forming ohmic contacts by the energy band alignment for *n*- and *p*-type 4*H*-SiC, respectively. Although these restrictions are less strict than for low doped semiconductors, there are no materials which can satisfy them. On the other hand, the ϕ_{Bn} and ϕ_{Bp} values are significantly lower than the one for contacts to low doped 4*H*-SiC. This should make formation of ohmic contacts through the carrier tunneling mechanism to heavily doped 4*H*-SiC much easier. Note that the values of ϕ_{Bn} and ϕ_{Bp} in metal-Si contacts are significantly lower than that one in metal-4*H*-SiC contacts for any value of metal work function. For the case of *p*-type Si, taking into account both Fermi level pinning and image-force barrier lowering results in increasing the choice of metals providing ohmic contacts to Si just by the energy band alignment.

The electrical current through the metal-semiconductor contact with Schottky barrier is determined by two different physical processes. The first one is a thermionic emission (TE) when the current is dominated by thermally excited carriers passing over the barrier. The second one is a field emission (FE) when the SCR is sufficiently narrow and charge carriers can tunnel through the barrier. In the intermediate case a thermionic-field emission (TFE) takes place when carriers are thermally excited to an energy where the barrier is sufficiently narrow for tunnelling [49]. Which process dominates at a specific doping level and temperature is defined by the ratio $qE_{00}/(k_BT)$ where k_BT is a thermal energy and E_{00} is defined as [10]:

$$E_{00} = \frac{\hbar}{2}\sqrt{\frac{N}{m_{tun}\varepsilon\varepsilon_0}} \tag{7}$$

where N is the semiconductor doping level, m_{tun} is the tunneling effective mass, and ε is the relative permittivity. The effective tunneling mass in a single valley is defined as [50]:

$$m_{tun} = \left(\frac{l_x^2}{m_x} + \frac{l_y^2}{m_y} + \frac{l_z^2}{m_z}\right)^{-1} \tag{7a}$$

where l_x, l_y, and l_z are the direction cosines of the carriers flux relative to the principle axes of the constant energy ellipsoid and m_x, m_y and m_z are the corresponding components of the effective mass tensor. The physical meaning of E_{00} is that it is the diffusion potential of a Schottky barrier (denoted in Fig. 1a as V_{bi}) at which the tunnelling probability for an electron at the bottom of the conduction band at the edge of the depletion region is equal to e^{-1} [10].

When $qE_{00}/(k_BT) \ll 1$ (usually this condition is replaced by strict inequality $qE_{00}/(k_BT) < 0.5$), TE is the dominating process and current densities through the Schottky barrier at forward and reverse biases, J_F and J_R, are described by [10, 12]:

$$J_F = J_S \exp\left(\frac{qV}{nk_BT}\right)\left(1 - \exp\left(-\frac{qV}{k_BT}\right)\right) \tag{8}$$

$$J_R \cong J_S = A^*T^2\exp\left(-\frac{\phi_B}{k_BT}\right) \tag{9}$$

where n is the ideality factor and A^* is the effective Richardson constant for electrons:

$$A^* = A_0\sum_{i=1}^{M_C}\frac{m_{ei}}{m_0} = \frac{qk_B^2m_0}{2\pi^2\hbar^3}\sum_{i=1}^{M_C}\frac{m_{ei}}{m_0} \tag{9a}$$

where A_0 is the Richardson constant for free electrons and m_{ei} is the electrons effective mass in the i-th single valley [20]. This effective mass is different from the effective tunneling mass defined by Eq. 7a and is given by:

$$m_{ei} = \sqrt{l_x^2 m_y m_z + l_y^2 m_x m_z + l_z^2 m_x m_y} \qquad (9b)$$

which can be rewritten for electrons in 4H-SiC as:

$$m_{ei} = \sqrt{l_{ML}^2 m_{e\,MK} m_{e\,M\Gamma} + l_{MK}^2 m_{e\,ML} m_{e\,M\Gamma} + l_{M\Gamma}^2 m_{e\,MK} m_{e\,ML}} \qquad (9c)$$

where $l_{ML,MK,M\Gamma}$ are directional cosines of the normal to the emitting plane relative to the principal axes of the ellipsoid.

Contact resistivity for TE can be easily derived from Eq. 8 as:

$$\rho_C = \left(\frac{\partial V}{\partial J}\right)_{V=0} = \frac{k_B T}{q A^* T^2} \exp\left(\frac{\phi_B}{k_B T}\right) \qquad (10)$$

Note that the contact resistivity does not depend on the ideality factor.

When $q E_{00}/(k_B T) \gg 1$ (usually replaced by $q E_{00}/(k_B T) > 5$) and
when $q E_{00}/(k_B T) \cong 1$ (usually replaced by $0.5 < q E_{00}/(k_B T) < 5$),
FE and TFE is the dominant process, respectively. The full theoretical expressions for the current density J and resistivity ρ_C at FE and TFE are complicated and can be found in [51]. For our estimations of contact resistivity, it is sufficient to know the functional dependence of contact resistivity on ϕ_B and N which is given (for n-type semiconductor) by [51]:

$$\rho_C \propto \exp\left(\frac{\phi_{Bn}}{\sqrt{N_D}}\right) \qquad \text{for FE } (q E_{00}/(k_B T) \gg 1) \qquad (11)$$

$$\rho_C \propto \exp\left(\frac{\phi_{Bn}}{\sqrt{N_D}\, \coth\left(\dfrac{q E_{00}}{k_B T}\right)}\right) \text{ for TFE } (q E_{00}/(k_B T) \sim 1) \qquad (12)$$

Indeed, as far as the contact resistivity at low doping levels is defined by TE mechanism and can be calculated exactly the following expression can be used to estimate contact resistivity in a full range of doping levels (for n-type semiconductor):

$$\rho_C = \frac{k_B T}{q A^* T^2} \exp\left(\frac{\phi_{Bn}}{k_B T}\right); \qquad \frac{q E_{00}}{k_B T} < 0.5$$

$$\rho_C \cong K_{TFE} \exp\left(\frac{\phi_{Bn}}{qE_{00} \coth\left(\dfrac{qE_{00}}{k_B T}\right)} \right); \qquad 0.5 < \frac{qE_{00}}{k_B T} < 5 \qquad (13)$$

$$\rho_C = K_{FE} \exp\left(\frac{\phi_{Bn}}{qE_{00}} \right); \qquad\qquad 5 < \frac{qE_{00}}{k_B T}$$

where the coefficients K_{FE} and K_{TFE} are used as a fitting parameters to stitch $\rho_C(N_D)$ in all three regions of N_D. Fig. 4 shows the contact resistivity depending on a semiconductor doping level calculated by Eq. 13 for Pt contact to p-type 4H-SiC (ϕ_{Bp}=1.19 eV at low doping); Ti contact to n-type 4H-SiC (ϕ_{Bn}=0.92 eV at low doping); and Ti contact to n-type Si (ϕ_{Bn}=0.56 eV at low doping). Both Fermi level pinning and barrier image-force lowering were taken into account. The contact metals were chosen from relatively chemically stable materials with lowest ϕ_M for n-type SiC and Si and with highest ϕ_M for p-type SiC. The $\rho_C(N_D)$ and $\rho_C(N_A)$ lines in Fig. 4 are drawn up to the corresponding impurity solubility limits and contact resistivities at these doping levels correspond to the theoretically estimated minimum available values of ρ_C. They are about:

$2 \times 10^{-7} \ \Omega \cdot cm^2$ for n-type 4H-SiC and

$8 \times 10^{-7} \ \Omega \cdot cm^2$ for p-type 4H-SiC

at doping levels close to the solubility limits which are ~$5 \times 10^{20} \ cm^{-3}$ for nitrogen in n-type 4H-SiC and ~$2 \times 10^{21} \ cm^{-3}$ for aluminum in p-type 4H-SiC. For comparison, the minimum resistivity of ohmic contacts to n-type Si according to these estimations is $6 \times 10^{-9} \ \Omega \cdot cm^2$.

Finally, it should be mentioned that ohmic contacts may be formed by creation of crystal defects near the semiconductor surface. These defects may act as recombination centers and lead to a significant reduction of contact resistance if their density is high enough. These defects may be created either before metal deposition by damaging the semiconductor surface or as a result of solid state chemical reaction between a metal and semiconductor during a post-deposition thermal treatment. To the best of our knowledge, there is no theory developed to make quantitative estimations of ρ_C value for contacts fabricated by this method.

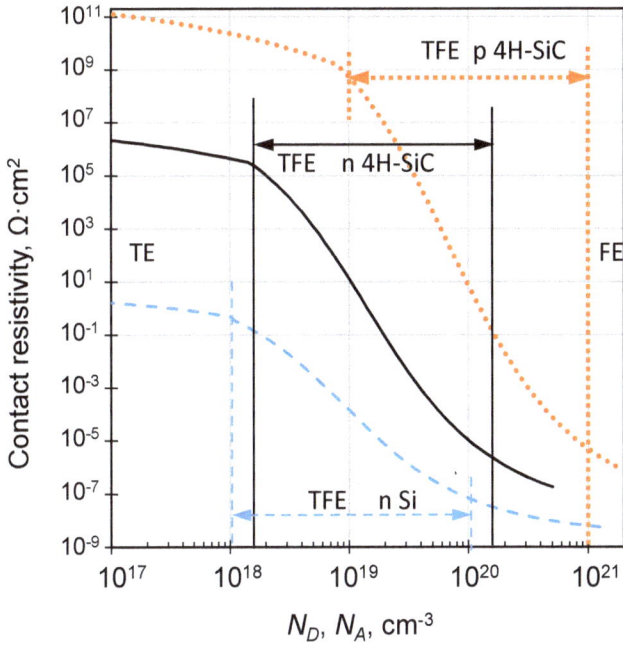

Figure 4. Contact resistivity depending on semiconductor doping level calculated by Eq. 13.
Dotted line: Pt contact to p-type 4H-SiC ($\phi_{Bp}=1.19$ eV at low doping); solid line: Ti contact to n-type 4H-SiC ($\phi_{Bn}=0.92$ eV at low doping); and dashed line: Ti contact to n-type Si ($\phi_{Bn}=0.56$ eV at low doping).

3. Methods, limitations and accuracy of contact resistivity measurements

A rough estimation of contact resistivity can be done by using a lateral two-terminal test structure shown in Fig. 5a. By neglecting a current non-uniformity and by a substrate resistance, the upper bound of contact resistivity is given by:

$$\rho_C = \frac{R}{2\pi r^2} \tag{14}$$

where r is the contact radius and R is the measured resistance. A vertical two-terminal test structure shown in Fig. 5b provides more accurate measurements taking into account the spreading resistance in the semiconductor. By neglecting the resistance of the bottom contact, the contact resistivity measured by this test structure is given by [52]:

$$\rho_C = \frac{1}{\pi r^2}\left(R - \frac{\rho_B}{2\pi r}\arctan\left(\frac{2x_S}{r}\right)\right) \tag{15}$$

where ρ_B is the semiconductor bulk resistivity and x_S is the semiconductor layer thickness. This method, sometimes called Cox and Strack method (CSM), requires knowledge of the substrate bulk resistance and is applicable only for contacts to uniformly doped substrates.

The resistivity of contacts to thin semiconductor layers can be measured by using cross bridge Kelvin resistance (CBKR) test structures, by a transfer length method (TLM) and by a circular modification of TLM (CTLM) [11]. The CBKR method is suitable for extraction of very low contact resistance but it requires separate test structures or CBKR test structures with multiple contacts for the measurement of the semiconductor sheet resistance. CBKR test structures are relatively complicated for fabrication. They require termination by mesa or local implantation and growth or deposition of a high quality insulation layer. This method was almost never used for resistivity measurements of contacts to silicon carbide. TLM provides direct measurements of the sheet resistance of thin semiconductor layers and is suitable for extraction of moderate and low contact specific resistances. TLM test structures require termination by mesa or by local implantation but do not require an insulation layer and their fabrication can be easily integrated in standard silicon carbide device processing. Circular modification of TLM does need a test structure termination and is easier to fabricate. CTLM provides measurements of contact resistivities in the same range as TLM but it is more space consuming than TLM and its accuracy is lower than TLM due to the larger size of test structures. Moreover, CTLM requires more complicated mathematics to extract a contact resistivity and the use of approximations further reduces its accuracy. As a result, TLM is the most preferred method to measure contact resistivity and almost all published results in ohmic contacts to SiC were obtained by TLM. Incorrect design of a TLM test structure and ignorance of measurements accuracy by this method may lead to significant errors in obtained values of contact resistivity. The following two sections briefly describe the transfer length method, test structure design, and provide quantitative estimations of TLM constraints and accuracy when it is applied for measurements of specific resistivity of contact to silicon carbide layers.

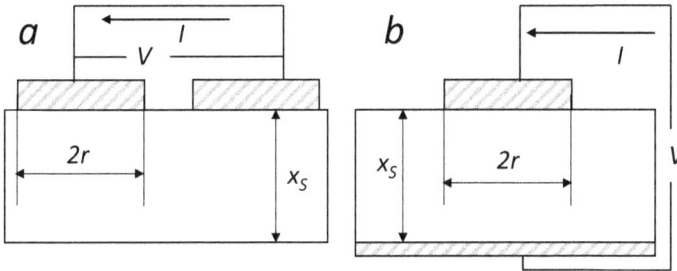

Figure 5.   *(a) A lateral two-terminal and (b) a vertical two-terminal test structure for measurements of contact resistance.*

3.1   Contact resistivity measurements by TLM

Fig. 6 shows a cross-section and a top view of a TLM test structure. It consists of $M+1$ contact pads of $L{\times}Z$ size which are located at unequal distances d_i of each other. The lateral current spreading has to be limited by a structure termination of width W. The semiconductor layer with the sheet resistance R_{Sh} has to be either insulated from the substrate by a p-n junction or to be formed on semi-insulating substrate. Following the TLM, the set of measured resistances between two adjacent contact pads for different pads separations, $R_i(d_i)$, are fitted to a linear relationship by the least-squares fitting procedure with equal weighting of the measured values. Then the intersections of the fitted line with R and d axes, $2R_{Cf}$ and L_y, are used to determine ρ_C and R_{Sh}. As an example, Fig. 7 shows measurements results $R_i(d_i)=V_i/I_i$ in a TLM test structure (Z=40 μm, W=48 μm, L=130 μm) formed on n-type 4H-SiC epi-layer (x_S=750 nm, N_D = $2.4{\times}10^{19}$ cm^{-3}) insulated by a p-n junction. Using multiple resistance measurements between contacts with different spacing in TLM allows us to extract the R_{Sh} value and the front contact resistance (R_{Cf}) from least squares fitting of $R_i(d_i)$ [53]:

$$R_i = \frac{V_i}{I_i} = Ad_i + B = \frac{R_{Sh}}{Z}d_i + 2R_{Cf} \tag{16a}$$

$$L_y = \frac{B}{A} = \frac{2ZR_{Cf}}{R_{Sh}} \tag{16b}$$

The current flow from the semiconductor layer in the metal contact is described by the transition line model [54]. The equivalent circuit of semiconductor-metal transition according to this model is shown in Fig. 8 and described by the set of equations:

$$\begin{cases} dI = -\dfrac{VZdy}{\rho_C} \\ dV = -\dfrac{IR_{Sk}dy}{Z} \end{cases} \tag{17}$$

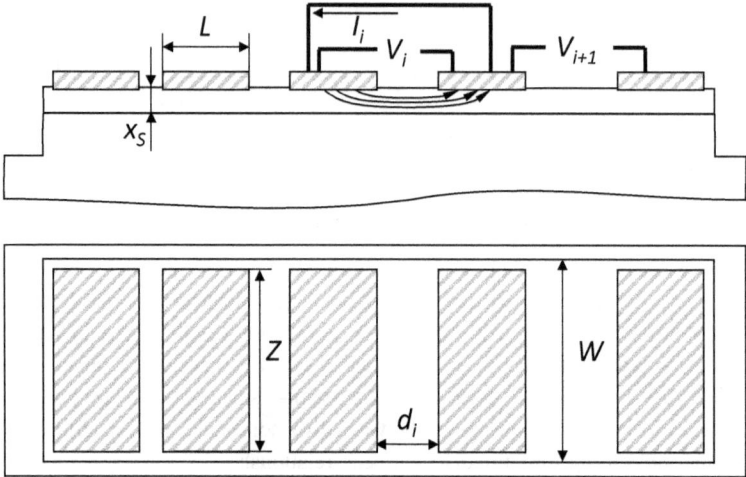

Figure 6. *Cross-section and top view of a TLM test structure for measurements of resistivity of contacts to thin semiconductor layers.*

where R_{Sk} is the semiconductor layer sheet resistance under the contact. This set of equations can be solved and the potential under the contact can be expressed in closed form [54]:

$$V(y) = \frac{I_i\sqrt{R_{Sk}\rho_C}}{Z} \frac{cosh\left(\frac{L-y}{L_T}\right)}{sinh\left(\frac{L}{L_T}\right)} \tag{18}$$

where

$$L_T = \sqrt{\rho_C/R_{Sk}} \tag{19}$$

is the transfer length – the distance from the contact edge where $V(L_T)=V(0)/e$. The L_T is considered as an effective contact length in the case of $L_T < L$ since the main part of the current flow is transferred from semiconductor to the metal contact at $y < L_T$. The front end contact resistance, R_{Cf} can be expressed as:

$$R_{Cf} = \frac{V(0)}{I(0)} = \frac{\sqrt{R_{Sk}\rho_C}}{Z} \coth\left(\frac{L}{L_T}\right) = \frac{\rho_C}{ZL_T} \coth\left(\frac{L}{L_T}\right) \qquad (20)$$

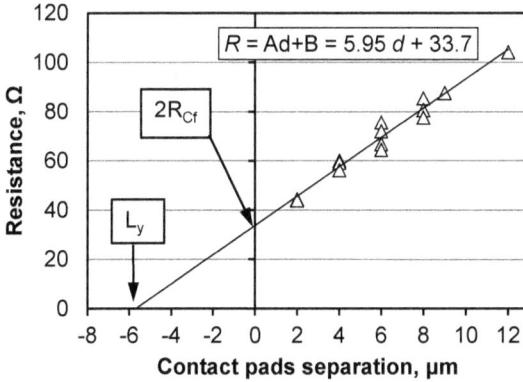

Figure 7. *Resistance measured between two adjacent contacts in a TLM test structure depending on contacts separation. The TLM test structure ($Z=40\,\mu m$, $W=48\,\mu m$, $L=130\,\mu m$) is formed on n-type 4H-SiC epi-layer ($x_S=750\,nm$, $N_D = 1.4\times10^{19}\,cm^{-3}$) insulated by a p-n junction model.*

Eq. 20 can be reduced to:

$$R_{Cf} \cong \frac{\rho_C}{ZL_T} \quad \text{when } L_T < L \qquad (21)$$

and

$$R_{Cf} \cong \frac{\rho_C}{ZL} \quad \text{when } L_T > L \qquad (22)$$

Usually contact pads have a reasonable size with both L and $Z \sim 100\,\mu m$ to place two probes and $L_T < L$ at all combinations of ρ_C and R_{Sh} in TLM test structures formed in n- and p-type 4H-SiC. Substituting Eq. 21 in Eq. 16b results in:

$$L_T = \frac{BZ}{2R_{Sk}} \qquad (23a)$$

$$\rho_C = \frac{Z^2 B^2}{4R_{Sk}} \qquad (23b)$$

$$R_{Sh} = AZ \qquad (23c)$$

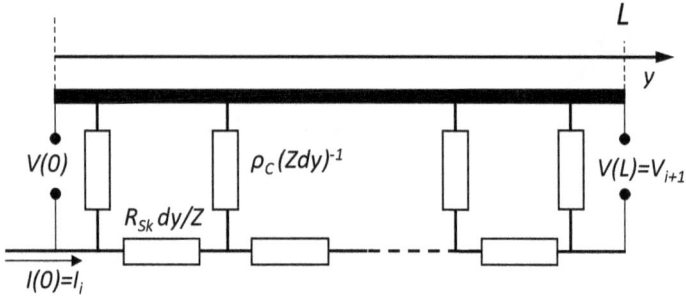

Figure 8. The equivalent circuit of current transition from the semiconductor layer in the metal contact according to the transition line model.

Eqs. 23 include an unknown semiconductor layer parameter – its sheet resistance under contact, R_{Sk}. In the first approximation, it may be assumed that $R_{Sk}=R_{Sh}$ and Eqs. 23 may be simplified to:

$$L_T = \frac{B}{2A} \tag{24a}$$

$$\rho_C = \frac{ZB^2}{4A} \tag{24b}$$

$$R_{Sh} = AZ \tag{24c}$$

Usually, $R_{Sk} > R_{sh}$ as a result of contacts post-deposition thermal processing and substituting $R_{Sk} = R_{Sh}$ in Eqs. 23 results in a conservative value of ρ_C calculated by Eq. 24b.

The value of R_{Sk} can be extracted from additional measurements of a contact end resistance R_{Ce} which is defined by the transition line model (see Fig. 8) as [53]:

$$R_{Ce} = \frac{V(L)}{I(0)} = \frac{\sqrt{R_{Sk}\rho_C}}{Z} \frac{1}{sinh\left(\frac{L}{L_T}\right)} = \frac{\rho_C}{ZL_T} \frac{1}{sinh\left(\frac{L}{L_T}\right)} \tag{26}$$

Combining Eq. 26 and Eq. 20 results in:

$$\frac{R_{Ce}}{R_{Cf}} = \frac{1}{cosh\left(\frac{L}{L_T}\right)} \tag{27a}$$

and

$$L_T = \frac{L}{arcosh\left(\frac{R_{Cf}}{R_{Ce}}\right)} \tag{27b}$$

This contact end resistance can be measured as it is shown in Fig. 6:

$$R_{Ce} = \frac{V(L)}{I(0)} = \frac{V_{i+1}}{I_i} \tag{28}$$

Substituting Eq. 19 and Eq. 23b in Eq. 27b allows to extract R_{Sk} and ρ_C:

$$R_{Sk} = \frac{ZB}{2L} arcosh\left(\frac{B}{2R_{Ce}}\right) \tag{29a}$$

$$\rho_C = \frac{LZB}{2 \, arcosh\left(\frac{B}{2R_{Ce}}\right)} \tag{29b}$$

Unfortunately, R_{Ce} is much lower than R_{Cf} and scarcely measurable with decent accuracy because of $L_T << L$ at almost all combinations of ρ_C and R_{Sh} in the TLM test structures formed in n- and p-type $4H$-SiC.

The R_{sk} value also can be estimated from analysis of contact processing parameters or from additional material and surface characterization. Indeed, the contact pads in the TLM structures used for measurements shown in Fig. 7, were formed by deposition of 80 nm thick Ni followed by rapid thermal processing (RTP) in vacuum at 1050 °C for 3 minutes. The values of $L_T = 2.8$ μm; $\rho_C = 1.9 \times 10^{-5}$ Ω·cm² and $R_{Sh} = 240$ Ω/sq. were calculated by Eqs. 24 for these measurements. The thickness of n^+ layer, x_{SiC}, in this structure was measured by the secondary ion mass spectrometry (SIMS) profiling. As a result of solid-state chemical reaction between Ni and SiC, all Ni was reacted with SiC to create Ni₂Si causing the n^+ layer thickness reduction under contact pads. The thickness of consumed SiC can be measured by surface profiling after stripping off the contacts or it can be calculated as:

$$\Delta x_{SiC} = \frac{N_{Ni}}{2N_{SiC}} x_{Ni} = 76 \, nm \tag{30}$$

where N_{Ni} and N_{SiC} is the atomic density of Ni and SiC, respectively. Then the sheet resistance under contact can be estimated as:

$$R_{Sk} = \frac{x_{SiC}}{x_{SiC} - \Delta x_{SiC}} R_{Sh} = 265 \, \Omega/sq \tag{31}$$

Then the values of $L_T = 2.5$ μm and $\rho_C = 1.7 \times 10^{-5}$ Ω·cm² can be calculated by Eqs. 23a and 23b.

Advancing Silicon Carbide Electronics Technology I   Materials Research Forum LLC
Materials Research Foundations **37** (2018)   doi: http://dx.doi.org/10.21741/9781945291852

3.2   TLM constraints

The transition line model used to extract contact resistivity from TLM measurements has two significant constraints: (1) the sheet resistance of the metal contact is considered to be negligibly low and (2) the current flow under the contact is considered to be essentially one-dimensional. The first condition is almost always satisfied for any solid and reasonably thick metal deposited on silicon carbide because of high SiC bulk resistivity. The second condition means that $x_S \ll L_T$ and should not be mixed with usually less strict constraint $x_S \ll d_i$. Let us assume that the upper borderline of x_S for correct use of the transition line model is $0.2 \cdot L_T$ [11]. Substituting $L_T = 5 \cdot x_S$ in Eq. 21 and 16b results in the value of $L_y = 10 \cdot x_S$ in Fig. 7 corresponding to this x_S borderline. This condition ($L_y > 10 \cdot x_S$) provides a fast and easy way to check a correct use of TLM by visual evaluation of $R_i(d_i)$ dependence shown in Fig. 7. The semiconductor layer thickness is not measured by TLM and furthermore, extraction of ρ_C value does not require the x_S value to be known. This leads to a common mistake when ρ_C is measured by TLM on very thick layers or even on bare substrates.

The $x_S \ll L_T$ condition also imposes the lower bound of contact resistivity which can be correctly measured by a TLM test structure formed on a semiconductor layer with specific thickness and bulk resistivity:

$$x_S \ll L_T = \sqrt{\rho_C/R_{Sk}} < \sqrt{\rho_C/R_{Sh}} = \sqrt{x_S \rho_C/\rho_B} \Rightarrow \qquad (32a)$$

$$\rho_C \gg \rho_B x_S \quad (\text{or } \rho_C = 25\, \rho_B x_S \text{ for } x_S = 0.2 L_T) \qquad (32b)$$

Eq. 32 takes into account that usually $R_{Sk} > R_{sh}$. Fig. 9 shows minimum contact resistivity which can be correctly measured by the TLM assuming that $x_S < 0.2 \cdot L_T$. There are three $\rho_{C|min}(x_S)$ lines shown for different bulk resistivities of semiconductor layers: $\rho_B = 100$ mΩ·cm which is about the minimum value reported for p-type 4H-SiC; $\rho_B = 10$ mΩ·cm which is about the minimum value reported for n-type 4H-SiC; and $\rho_B = 0.5$ mΩ·cm which is about the minimum value reported for n-type Si, for comparison. For example, the TLM test structures used in measurements shown in Fig. 7 were formed on 0.75 µm thick n-type 4H-SiC layer and hence should be used for measurements of contact resistivity not lower than 2×10^{-5} Ω·cm^2 at any available doping level according to Fig. 9. In the strict sense, the ρ_C value of 1.7×10^{-5} Ω·cm^2 extracted by TLM on that test structures was not correct and thinner semiconductor layer should be used for that measurements. Note that the resistivity below 10^{-5} and 10^{-6} Ω·cm^2 of contacts to p- and n-type 4H-SiC, respectively, can be correctly measured by TLM only on 4H-SiC layers with $x_S < 40$ nm and this constraint is independent on a contact pad size and spacing. 4H-SiC layers with the thickness of 40 nm are too thin for practical use since ohmic contacts usually require post-deposition thermal processing which may lead to consumption of a significant part of a SiC layer comparable with its thickness. For comparison, the resistivity of about $10^{-7}\,\Omega\cdot cm^2$ can be correctly measured by TLM in contacts to reasonably thick n-type Si layers ($\sim$100 nm) because of much lower bulk resistivity of silicon.

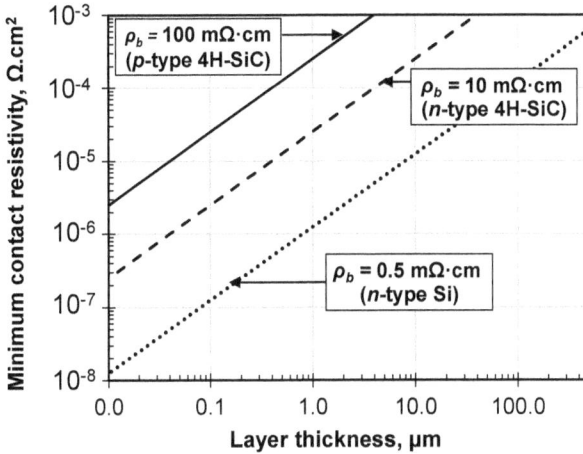

Figure 9.   Minimum contact resistivity which can be correctly measured by TLM on semiconductor layers with bulk resistivities of 100, 10 and 0.5 mΩ·cm (assuming that the transmission line model is applicable at $x_S < 0.2 \cdot L_T$).

3.3   TLM accuracy

There are three parameters contributing to the uncertainty of extracted contact resistivity by TLM. They are errors of measured resistances, of contact pads separations and of contact pads widths. These errors can be either random or systematic. The analytical expressions for total relative uncertainties of contact resistivity and of sheet resistance were derived by Ueng *et al.* [55] in the assumption that $R_{sk}=R_{sh}$ and that d_i are uniformly spaced: $d_i=id_{max}/M$ where $i = 1...M$, and M is the number of contact pad spacings, $M>>1$.

For random errors, the uncertainty is determined by the accuracy of used measurement equipment, by material defects, by fluctuations of contact pads geometry and by fluctuations in measurements conditions. The total relative uncertainty of contact resistivity and sheet resistance by random error can be expressed as [55]:

$$\frac{\sigma \rho_C}{\rho_C} = \frac{1}{\sqrt{M}}\left(\left(\frac{2Z}{\sqrt{\rho_C R_{Sh}}} + \frac{2\sqrt{3}Z}{R_{Sh}d_{max}}\right)\sqrt{\left(\frac{R_{Sh}}{Z}\right)^2 \cdot (\sigma d)^2 + (\sigma R)^2 + 4\left(\frac{\sigma Z}{Z}\right)}\right) \quad (33)$$

$$\frac{\sigma R_{Sh}}{R_{Sh}} = \frac{1}{\sqrt{M}}\left(\left(\frac{2\sqrt{3}Z}{R_{Sh}d_{max}}\right)\sqrt{\left(\frac{R_{Sh}}{Z}\right)^2 (\sigma d)^2 + (\sigma R)^2 + 2\left(\frac{\sigma Z}{Z}\right)}\right) \quad (34)$$

where σ denotes the standard deviation of measured parameters. These equations can be simplified for $L_T << d_{max}$:

$$\frac{\sigma \rho_C}{\rho_C} = \frac{2}{\sqrt{M}}\left(\sqrt{\left(\frac{\sigma d}{L_T}\right)^2 + \left(\frac{\sigma R}{R_{Cf}}\right)^2 + 2\left(\frac{\sigma Z}{Z}\right)}\right) \quad (33a)$$

$$\frac{\sigma R_{Sh}}{R_{Sh}} = \frac{1}{\sqrt{M}}\left(2\sqrt{3}\sqrt{\left(\frac{\sigma d}{d_{max}}\right)^2 + \left(\frac{\sigma R}{R_{max}}\right)^2 + 2\left(\frac{\sigma Z}{Z}\right)}\right) \quad (34a)$$

where $R_{max} = \frac{R_{Sh}d_{max}}{Z}$. The σd and σZ values are defined by the tolerance of used photomask (usually about 0.1 μm) and by the photolithography reproducibility through the processed wafer. In following estimations, the σd and σZ values are assumed to be about 0.1 μm. The σR value is mainly defined by the accuracy of measurement equipment. For example, a commonly used semiconductor device analyzer (Keysight B1500A with medium power SMU) provides the measurement accuracy about 2 mV in the ± 5 V range ($\sigma V/V \approx 4 \times 10^{-4}$) and 2.5 μA in the ±10 mA range ($\sigma I/I \approx 2.5 \times 10^{-4}$). These values result in $\sigma R/R \approx 6.5 \times 10^{-4}$. The maximum measured resistance usually is about 5 kΩ and 500 Ω for p- and n-type 4H-SiC layers, respectively, resulting in $\sigma R \approx 3$ Ω for measurements on p-type 4H-SiC layers and $\sigma R \approx 0.3$ Ω for measurements on n-type 4H-SiC layers. These values will be used in following estimations. Increasing the M number leads to noticeable reduction of measurements uncertainty due to random error. Theoretically, it can be chosen sufficiently large to reduce the relative random error of measured contact resistivity and sheet resistance below any predetermined value. In practice, the M number is limited by usable space and usually does not exceed 10. A TLM test structure with 10 contact pads ($M=9$) will be used in following estimations.

The uncertainty due to the systematic error is more significant in the TLM measurements. The TLM technique involves least-squares fitting $R_i(d_i)$ by a linear relationship and extracting the $2R_{Cf}$ value intercepted by the fitted line on the vertical axis. In contrast to the random error which appears as randomly scattered measured points around an average value and does not change the average $2R_{Cf}$ value, the systematic error appears as

Advancing Silicon Carbide Electronics Technology I   Materials Research Forum LLC
Materials Research Foundations **37** (2018)   doi: http://dx.doi.org/10.21741/9781945291852

a shift of the whole least-square fitted line with corresponding change of the $2R_{Cf}$ value because of all measured points have the same δd and/or δR values. The total relative uncertainties of contact resistivity and sheet resistance due to systematic error are [55]:

$$\frac{\delta \rho_C}{\rho_C} = \left(\frac{Z}{\sqrt{\rho_C R_{Sh}}}\right)\delta R + \left(\sqrt{\frac{R_{Sh}}{\rho_C}}\right)\delta d + 4\left(\frac{\delta Z}{Z}\right) \tag{35}$$

$$\frac{\delta R_{Sh}}{R_{Sh}} = 2\left(\frac{\delta Z}{Z}\right) \tag{36}$$

where δ denotes the systematic error. Eq. 35 may be rewritten as:

$$\frac{\delta \rho_C}{\rho_C} = \frac{\delta R}{R_{Cf}} + \frac{\delta d}{L_T} + 4\left(\frac{\delta Z}{Z}\right) \tag{35a}$$

For systematic error, the δZ is defined by the width of test structure termination, W. It is assumed that W-Z<<Z, but for conventional contact photolithography available in a lab, the realistic estimate is W-Z≈4 μm. Effective contact width is somewhere between Z and W due to current spreading and it can be reasonably assumed that δZ ≈ 2 μm for use in following estimations.

Systematic error of contact pads separations, δd, is defined by the contact patterning process. Usually the lift-off procedure is used which provides a geometry control with better precision than direct metal etching and can be easily used for patterning multilayer films. A negative or reversible photoresist used for the lift-off procedure has to be much thicker than the deposited film and usually is about 1.5 μm. Very common sources of systematic error in contact pads separations are photoresist under-edge exposure due to bad contact with photomask, photoresist over/under exposure, photoresist undercut due to overdevelopment, photoresist reflow during the post bake. Control of d_i during processing is usually performed by an optical microscope and δd≈0.2 μm is a very optimistic estimate for this kind of measurements. Unfortunately, this control is omitted quite often and d_i values given in a photomask are used resulting in significantly higher δd values.

The main contributions to δR are the resistances of probes, wires and cable connections which can be about several ohms in total. The 4-probe resistance measurements configuration where the split between voltage and current leads is realized at a contact pad has to be always used to minimise this error. A direct indication about using this configuration has to be given in a publication to convince that this issue was addressed.

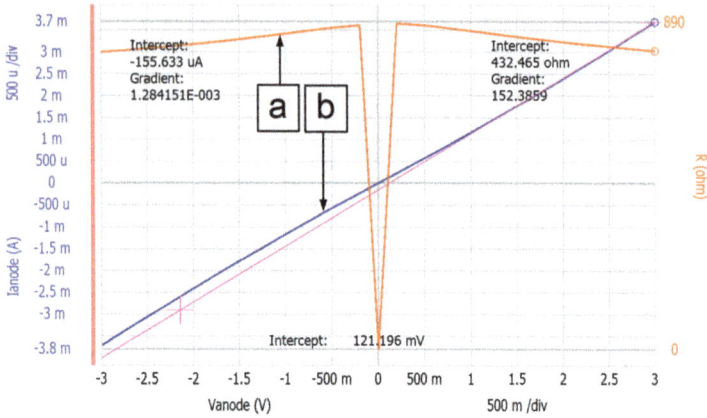

Figure 10.   I-V characteristic measured between two contact pads with 10 µm spacing in a TLM test structure (line "b") and differential resistance depending on voltage bias (line "a"). The TLM test structure (Z=160 µm, W=170 µm, L=100 µm) is formed on a p-type 4H-SiC epi-layer (x_S=170 nm, N_A = 2×10^{19} cm^{-3}).

Another source of systematic resistance variations is its incorrect extraction from measured *I-V* characteristics. Indeed, ohmic contact definition does not require it to demonstrate an ohmic behaviour and its *I-V* characteristics may be "quasi-linear". There is a precise mathematical definition of the graph curvature but it is never used as quantitative criteria of *I-V* characteristics linearity in publications. Instead, a typical *I-V* characteristic measured between two adjacent contacts may be provided and the estimation of its degree of linearity is entirely based on visual evaluations. This may lead to an incorrect use of static resistance instead of differential resistance at zero bias which is required by definition of contact resistivity. For instance, the curve (b) in Fig. 10 shows *I-V* characteristic measured between two adjacent contacts in a TLM test structure and the curve (a) shows a differential resistance depending on voltage bias. The TLM test structure (Z=160 µm, W=170 µm, L=130 µm) was formed on a *p*-type 4H-SiC epi-layer (x_S=170 nm, N_A = 2×10^{19} cm^{-3}). Measured *I-V* characteristic is quasi-linear and the difference $dV/dI|_{V=0}$ -V/I is about 76 Ω. Obviously, this difference does not depend on contact pad separations and is visually undetectable in *I-V* characteristics measured between contacts with larger separation distances.

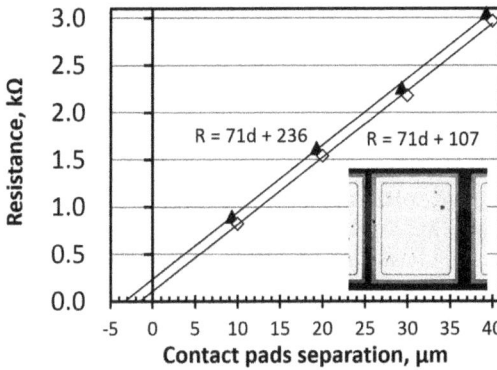

Figure 11.   Total resistances acquired for the different pad spacings a TLM test structure structure (Z=160 μm, W=170 μm, L=100 μm) formed on a p-type 4H-SiC epilayer (x$_S$=170 nm, N$_A$ = 2×10^{19} cm^{-3}).
Open diamonds – static resistance depending on incorrect distances. Solid triangles – differential resistance at zero voltage bias depending on corrected distances between contact pads.

Fig. 11 shows the effect of systematic errors δd and δR on the value of extracted contact resistivity in this particular TLM test structure (shown in the inset). Open diamonds represent measurements results where R was calculated as a static resistance and d_i values were taken from the photomask drawing. Corresponding contact resistivity of 6.5×10^{-5} $\Omega \cdot$cm^2 was extracted from this measurements. An inspection of this TLM test structure under optical microscope revealed that all d_i values are reduced for $\delta d \approx 0.7$ μm due to photoresist undercut. This value and $\delta R = 76$ Ω were used to get a correct $R(d_i)$ dependence (shown by solid triangles) and corresponding contact resistivity of 3.1×10^{-4} $\Omega \cdot$cm^2 was extracted. Note that the relative systematic errors $\delta d/d < 7\%$ and $\delta R/R < 8\%$ resulted in $\delta \rho_C/\rho_C \approx 80\%$. If the 4-probe configuration is used and differential resistance is correctly calculated, the systematic error can be assumed to be equal to the random error for measured resistances. This assumption will be used in following estimations.

When the values of δR and δZ are estimated, the contact pad width yielding the minimum systematic error of contact resistivity, Z_{opt}, can be found by evaluating the partial derivative of Eq. 35 with respect to Z and setting it equal to zero [55]:

$$Z_{opt} = 2\sqrt{L_T \delta Z \frac{R_{Sh}}{\delta R}} \qquad (37)$$

Advancing Silicon Carbide Electronics Technology I   Materials Research Forum LLC
Materials Research Foundations **37** (2018)   doi: http://dx.doi.org/10.21741/9781945291852

Further optimisation of a TLM test structure to minimize the uncertainty in extracted contact resistivity can be done by corresponding choice of contact separation distances. First, it should be noted that the ratios $\sigma R/R_{Cf}$ and $\sigma d/L_T$ appear in the Eq. 33a instead of relative errors $\sigma R/R$ and $\sigma d/d$. Similar, the ratios $\delta R/R_{Cf}$ and $\delta d/L_T$ appear in the Eq. 35a instead of relative errors $\delta R/R$ and $\delta d/d$. The value of R_{Cf} is always smaller than measured total resistances and to avoid a significant increase of the relative error in front contact resistance, the $2R_{Cf}$ value intercepted by the fitted line on the vertical axis and the maximum measured total resistance, R_{max}, should be in the same measurement range of used test equipment. This leads to a simple limitation on d_{max}:

$$R_{max} \leq 10 \times \left(2R_{Cf}\right) \Rightarrow 2R_{Cf} + \frac{d_{max}R_{Sh}}{Z} \leq 10R_{Cf} \Rightarrow d_{max} \leq 18L_T \qquad (38)$$

3.4   Practical example of a TLM test structure design and parameters calculation

Summarizing the above discussions on the TLM constrains and accuracy, Table 3 shows an example of geometry design and parameters calculation for a TLM test structure formed on 4H-SiC with minimum available bulk resistivity. The set of initial data includes SiC bulk resistivity, projected layer thickness and uncertainties of measurements and processing estimated above.

Fig. 12 shows resulted random and systematic relative errors of extracted contact resistivity calculated by Eq. 33a and Eq. 35a, respectively, for the TLM test structure with parameters described in Table 3. Note that the total error of extracted contact resistivity approaches 50% at the minimum measurable value of ρ_C.

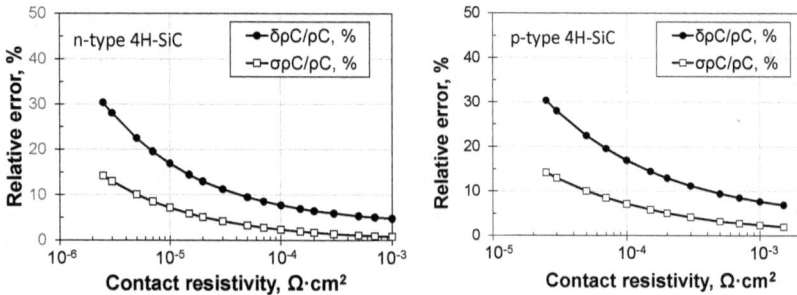

Figure 12.   Relative uncertainty of extracted contact resistivity due to systematic and random errors in resistance measurements and definition of contact pads width and separation distances in the TLM test structure with parameters described in Table 3.

These error estimations and calculations of optimized TLM test structure parameters were made for the lowest available SiC bulk resistivity, measuring accuracy of a standard test equipment and patterning tolerance provided by commonly used conventional contact photolithography. Consequently, it may be concluded that the minimum contact resistivity which can be measured with decent accuracy by TLM is 2.5×10^{-6} $\Omega \cdot cm^2$ and 2.5×10^{-5} $\Omega \cdot cm^2$ for ohmic contacts formed on n- and p-type 4H-SiC, respectively. Lower reported ρ_C values should be considered as a rough estimation unless detailed analysis of measurements constrains and errors is provided.

Table 3. Calculation of TLM test structure parameters.

		n 4H-SiC	p 4H-SiC
Input data:			
Minimum available bulk resistivity, ρ_B	$\Omega \cdot cm$	0.01	0.1
Thickness of epi-layer or implantation depth, x_s	nm	100	100
Layer sheet resistance, $R_{sh} = \rho_B/x_s$	Ω/sq.	1000	10000
Number of measurements, M		9	9
Systematic error of contacts widths, δZ	μm	2	2
Standard deviation of contacts widths, σZ	μm	0.1	0.1
Accuracy of resistance measurements, $\sigma R = \delta R$	Ω	0.3	3
Accuracy of contact spacing measurements, $\sigma d = \delta d$	μm	0.1	0.1
Calculated parameters:			
Minimum measurable ρ_C, calculated by Eq. 32b	$\Omega \cdot cm^2$	2.5×10^{-6}	2.5×10^{-5}
Transfer length L_T calculated by Eq. 19	μm	0.50	0.50
Max. contact separation distance, d_{max} calculated by Eq. 38	μm	9	9
Z_{opt}, calculated by Eq. 37	μm	115	115
R_{cf} calculated by Eq. 21	Ω	4.3	43
Minimum contact separation distance, d_{max}/M	μm	1	1
R_{max} calculated by Eq. 38	Ω	86.6	866

4. Fabrication of ohmic contacts to n-type SiC

In a laboratory practice, acceptable ohmic contacts to silicon carbide with n-type of conductivity can be easily fabricated by electrical discharge machining (EDM). The process is illustrated by Fig. 13. A simple capacitor discharge circuit or any convenient laboratory DC power supply can be used to produce a spark. The Aim-TTi EX752M 75 V/2 A PSU was used in this particular example. This PSU has a 200 µF output capacitor providing enough energy to produce a spark. The procedure to form a crater in the SiC surface is as following: set the current limit to 10 mA and output voltage about

30 V, place one contact firmly on the SiC wafer and approach the second contact (a tungsten probe station needle) until a spark ignites (Fig. 13a) and produces a single black spot on the SiC surface (Fig. 13b). A required voltage depends on wafer conductivity and surface state and may need to be adjusted. A single spark produces a crater in the SiC surface having very rough bottom surface and filled with soot (Fig. 13c).

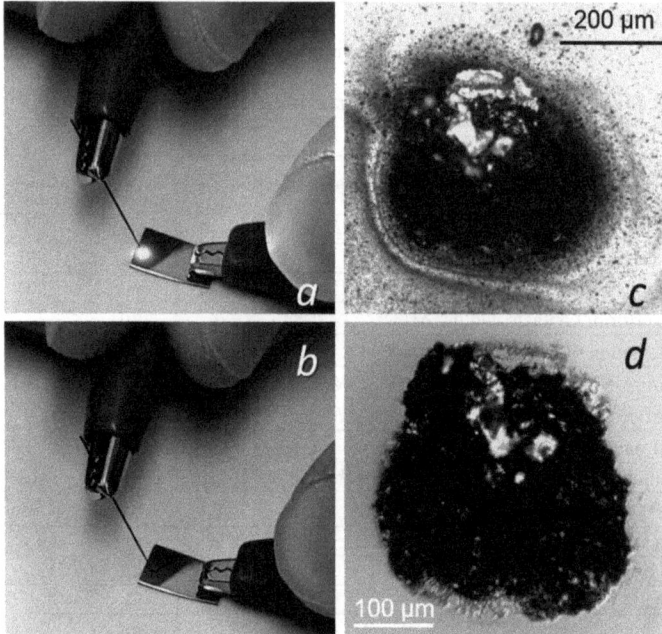

Figure 13. Fabrication of ohmic contacts to silicon carbide with n-type of conductivity by electrical discharge machining.

Finally, any soft metal (indium in this instance) has to be spread over the crater to form an ohmic contact as it is shown in the inset in Fig. 14. *I-V* and R_d-*V* characteristics measured in the configuration with two top contacts formed by this procedure are shown in Fig. 14. The contacts were formed on a commercial 4*H*-SiC wafer with ρ_B=0.02 Ω·cm. They demonstrated almost linear *I-V* characteristic. To estimate the contact resistivity, the metal overlays were pilled off after measurements and the wafer was cleaned by wet chemistry to remove metal traces and by oxygen plasma using Tegal PLASMOD 100 W Tabletop Plasma Reactor to remove carbon. Fig. 13d shows the crater in the SiC surface after metal and soot removal. Measured crater diameter of about 300 μm and $R_d(0)$=24 Ω extracted from Fig. 14 give a rough estimation for the upper bound of contact resistivity $\rho_C < 5 \times 10^{-3}$ Ω·cm².

This is a very simple and efficient laboratory method to form ohmic contacts with decent resistivity to SiC samples with *n*-type of conductivity. It has been used since the early days of SiC technology and tested on both 6*H* and 4*H* commercial wafers and Lely crystals with moderate and high doping. For example, this method was used to form backside contacts in 6*H*-SiC dinistors operating at temperatures up to 500 °C [56]. Unfortunately, it is not compatible with SiC device processing when ohmic contacts are required to thin epitaxial layers since EDM destroys material for a significant depth usually exceeding thickness of a device structure, cannot be controlled with acceptable accuracy and produced contacts are not usable at0 elevated temperatures. It still remains very usable for quick and easy fabrication of back side contacts in SiC test structures for measurements at temperatures ranged from cryogenic to moderately high.

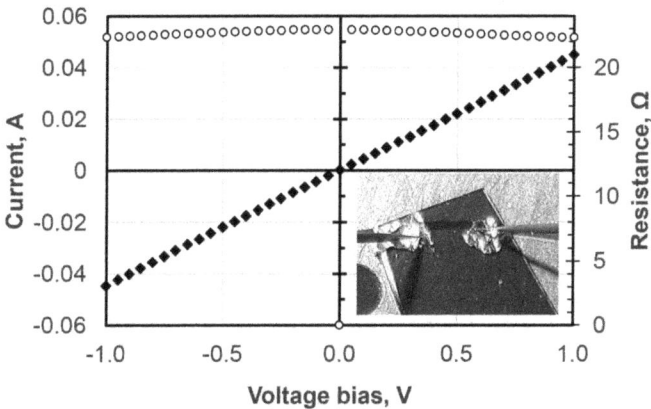

Figure 14.   I-V characteristic (solid diamonds) and differential resistance depending on voltage bias (open circles) measured in two ohmic contacts formed on n-type 4H-SiC wafer by EDM.

Conventional methods of ohmic contacts fabrication compatible with SiC device processing include growth or deposition of contact materials followed by optional thermal treatment. After thermal processing, contacts have to be covered by a cupping

Advancing Silicon Carbide Electronics Technology I   Materials Research Forum LLC
Materials Research Foundations **37** (2018)   doi: http://dx.doi.org/10.21741/9781945291852

layer of highly conductive soft metal for further probe testing or wire bonding. They have to be protected from oxidation at high temperatures which potentially may exceed 700 °C for SiC devices. Table 4 and Table 5 list ohmic contacts parameters, characterization and fabrication details reported for *n*-type 4*H*- and 6*H*-SiC, respectively. The data presented in these tables were taken from [57], [8] and [9], critically revised (some controversial data were excluded) and updated with recently published results. Contact resistivities listed in these tables were measured by TLM if another method is not indicated in comments.

Table 4 Ohmic contacts to n-type 4H-SiC.

Metals*	Thick ness (nm)	SiC layer info, x_S (nm)	$N_D/10^{18}$ (cm^{-3})	Annealing Conditions			$\rho_c/10^{-5}$ ($\Omega\cdot$cm^2)	surface treatment, deposition method, comments	Ref.
				T (°C)	t (min)	ambient			
as deposited									
Ti-C	150 co-sptr. at 500 °C	1000 epi	13	0	0		0.928	UHV; TiC epi	[58]
Al	50	200 P impl in 300 epi	500	0	0		0.054	sacr.ox.	[59]
Al	n/r	200 P impl in 300 epi	500	0	0		0.12	sacr.ox.	[60]
Mo	50	200 P impl in 300 epi	500	0	0		0.2	sacr.ox.	[59]
Ni	50	200 P impl in 300 epi	500	0	0		0.3	sacr.ox.	[59]
Ti	50	200 P impl in 300 epi	500	0	0		0.027	sacr.ox.	[59]
Ni based									
Ni	100 e-beam	x_S n/r; P impl at 400 °C	300	1000	1	N$_2$	10		[61]
Ni	50	800 epi	15	1000	2	Ar	0.033		[62]
Ni	50	200 P impl in 300 *n*-epi	500	1000	2	Ar	0.22	sacr.ox.	[60]
Ni	50	x_s n/r; P impl at 700 °C	250	1000	2	Ar	0.04		[63]
Ni	50	epi	20	1100	2	Ar	12		[64]
Ni	170 e-beam	250; N impl	18	1040	3	vac	1.5	sacr. ox.	[65]
Ni	50-500	200; epi	5	1000	5	n/r	2		[66]
Ni	150 e-beam	C-face; substr.	3	950	5	1% H$_2$/Ar	0.49	Ar+ ions	[67]
Ni	150 e-beam	200; epi on 500 epi	10	950	10	N$_2$	0.28		[68]

| Metals* | Thick ness (nm) | SiC layer info, x_S (nm) | $N_D/10^{18}$ (cm^{-3}) | Annealing Conditions | | | $\rho_C/10^{-5}$ ($\Omega \cdot$cm^2) | surface treatment, deposition method, comments | Ref. |
				T (°C)	t (min)	ambient			
Ni/TaSi$_x$/Pt	100/200/400; sptr.	1000; epi	11	950	30	Ar	35	sacr. ox.	[69]
Ni	50; sptr	x_S n/r; N impl	30	950	n/r	Ar	21		[70]
Ni	50; therm	x_S n/r; epi	13	1000	5	UHV	18	CTLM	[71]
Ni/Si	150 e-beam	200; epi on 500 low doped n-epi	10	950	10	Ar	2.7	Ni/Si ratio stoichiometric to NiSi$_2$	[68]
Ni/Si	150 e-beam	200; epi on 500 low doped n-epi	10	950	10	N$_2$	0.27		[72]
Ni/Si/Ni	100/50/150; sptr	10000; epi on n-substr	11	550+800	10+3	10%H$_2$/Ar	1.4		[73]
Ni/TiN	50/10	100 N impl	1000	1000	0.5	Ar	3.5	Ar ICP, CTLM by direct met. etch., N_D> solubility limit	[74]
Ni/TiW	100/200; sptr	C-face substr	5	975	1	N$_2$	4.2	CTLM	[75]
Ni80Cr20	x_m n/r; sptr	epi	50	1000	2	vac	0.5		[76]
Ni80Cr20	50, 200 sptr	500 epi	13	1100	3	vac	1.2		[77]
Ti based									
Ti	50	epi	20	1100	2	Ar	12		[64]
Ti/TaSi$_2$/Pt	100/400/200;sptr	2000; epi	20	600	30	N$_2$	47	four-point probe	[78]
Ti	30; sptr	x_S n/r; N impl	30	1050	n/r	Ar	65		[70]
Ti/Al/Si	20/30/30	x_S n/r; impl	26	1020	2.5	Ar	0.37	sacr. ox.	[79]
Ti/Ni	10/20; e-beam	230; N impl	100	950	1	N$_2$	0.48		[80]
Ti/Ni	10/20; e-beam	500; N impl	10	1000	10	Ar	4.1	CBKR	[81]
Ti/TiN/Al	100/10/300	100 P impl	1000	600	5	vac	0.083	Ar ICP, CTLM, direct met. etch.	[74]
Ti$_3$SiC$_2$	co-sptr. at 800 °C	800 epi	15	950	1	Ar	50		[82]
Ti-C	150 co-sptr. at 500 °C	1000 epi	13	950	2	10%H$_2$/Ar	4.0	UHV; TiC epi	[58]
Ti(30 wt.%)W	110; sptr	200; epi on 400 n-epi	50	950	5	Ar	40.8		[83]
TiW	180 sptr. at 200 °C	1000 epi	11	950	30	vac	3.3	ICP etched	[84]
TiW	180 sptr.	1000 epi	11	950	30	Ar	1.5	sacr. ox.	[69]

Metals*	Thickness (nm)	SiC layer info, x_S (nm)	$N_D/10^{18}$ (cm^{-3})	Annealing Conditions			$\rho_C/10^{-5}$ ($\Omega\cdot$cm^2)	surface treatment, deposition method, comments	Ref.
				T (°C)	t (min)	ambient			
Other metals									
Al/Ni	~6%/50 e-beam	200; P impl	200	1000	2	Ar	4.8		[85]
Al/Ti/Au	150/150/100 therm	substr, no epi	10	1050	5	Ar	0.28	CTLM	[86]
Al/Ti/Ni	50/50/20 e-beam	x_s n/r; epi	10	800	30	UHV	20	sacr. ox. CTLM	[87]
Nb	200	500 epi	13	1100	10	n/r	0.1	ρ_C limited by TLM accuracy	[88]
WNi/Si	100/20 W - 75 at.%	2000-3000 epi	6	1100	60	Ar	50		[89]
TaSi$_x$/Pt	200/400 sptr.	1000 epi	11	950	10	Ar	1	sacr. ox.	[69]

* In multilayer contacts, the metal on the left hand side was deposited first.
Abbreviations: n/r: not reported; e-beam: e-beam evaporation; therm.: thermal evaporation; sptr.: magnetron spattering; Ti/Al: layered deposition; Ti-Al: co-deposition from two sources; TiAl – alloy sputtering; epi: epitaxial layer; impl: implanted; vac: vacuum; UHV: ultrahigh vacuum (~3×10^{-10} Torr).

The first surprising conclusion from the analysis of data collected in Table 4 and Table 5 is apparent absence of significant progress in reducing resistivity of contacts to *n*-type SiC. Indeed, Fig. 15 shows reported specific resistances of ohmic contacts to *n*-type SiC depending on a publication date with trendlines shown by dotted lines. There may be various speculation to explain this tendency, but it should be noted that the fabrication technology and characterization methodology of ohmic contacts were sufficiently matured to the time when development of low resistivity ohmic contacts to SiC became a topical problem. The experience gained in silicon and A3B5 materials provided a good starting point to get low resistivity contacts very soon after starting these research in SiC. Slight increase in published ρ_C values during following years may be associated with significant progress in SiC material growth, epitaxy and ion implantation because of as better is the crystal quality of underlying material then more complicated is fabrication of ohmic contacts to it.

Table 5 Ohmic contacts to n-type 6H-SiC.

Metals*	Thickness (nm)	SiC layer info, x_S (nm)	$N_D/10^{18}$ (cm^{-3})	Annealing Conditions			$\rho_C/10^{-5}$ ($\Omega\cdot$cm^2)	surface treatment, deposition method, comments	Ref.
				T (°C)	t (min)	ambient			
Ni based									
Ni	100	substr., no epi	7.4	950	1	N$_2$	3.9		[90]
Ni	x n/r; e-beam	500; epi	8	950	2	vac	0.5		[91]

Metals*	Thickness (nm)	SiC layer info, x_S (nm)	N_D /10^{18} (cm^{-3})	Annealing Conditions			ρ_C/10^{-5} ($\Omega \cdot cm^2$)	surface treatment, deposition method, comments	Ref.
				T (°C)	t (min)	ambient			
Ni	x n/r; e-beam	Si, C Lely	2	1000	2	vac	8		[92]
Ni	x n/r; e-beam	Si, C Lely	5	1000	2	vac	8		[93]
Ni	200	C, LPE on Acheson	450	1000	5	Ar	0.1	CSM	[94]
Ni	50-500	Si	20	1000	5	N_2	0.2		[66]
Ni	100	substr	1	1020	5	N_2:H_2 (99:1)	21	Shadow mask; four-point probe	[95]
Ni	200; e-beam	C; substr	1	1000	5	Ar	300		[96]
Ni	100	substr	7.4	950	n/r	N_2	3.9		[97]
Ni	100; e-beam	500 epi	8.1	950	5	vac	5.2		[98]
Ni	150 e-beam	substr., no epi	1.8	950	10	Ar	1		[68]
Ni/Si	150 e-beam	substr., no epi	1.8	950	10	Ar	30	Ni/Si ratio stoichiometric to NiSi$_2$	[68]
Ni80Cr20	50, 200 sptr	500 epi	1.4	1100	3	vac	9.1		[77]
Si/Ni	50/n.r LPCVD/sptr.	Epi	15	300	540	N_2	69	CTLM	[99]
Ti based									
Ti	100 e-beam	substr., no epi	7.4	900	n/r	vac	10	sacr. ox.; TiC/Ti$_5$Si$_3$	[100]
Ti	100 e-beam	substr., no epi	7.4	1000	n/r	vac	6.7	sacr. ox.; Ti$_3$SiC$_2$	[100]
Ti/Ni	20/120	on-axis subst., no epi	2	950	2	Ar	5.9		[101]
Ti/TaSi$_2$/Pt	100/400/200; sptr	2000; epi	7	600	30	N_2	16.8	sacr. ox.; four-point probe	[78]
TiC	150 CVD	1000; epi	40	1300	15	N_2	1.3		[102]
TiSi$_x$	400; sptr	300; N impl	5	1150	2	Ar	0.7		[103]
Other metals									
Ta/Ni/Ta	20/70/10; sptr	substr	0.6	800	10	Ar	30		[104]
Re	100	substr., no epi	1.28	1000	120	vac	7	TLM by shadow mask	[105]
MoSi$_2$	100 sptr	1000; epi	10	1000	20	H_2	5.2	PECVD SiO$_2$/BOE	[106]
Nb	200	500 epi	1.4	1100	10	n/r	0.3		[88]
TaC	160-320 sptr. at 200 °C	substr., no epi	23	1000	15	vac	2.1	550C/10 min vac	[107]
TaC	100 sptr. at	500 epi on n-	8	1000	15	vac	3	550C/10 min vac;	[108]

Metals*	Thick ness (nm)	SiC layer info. x_S (nm)	N_D /10^{18} (cm^{-3})	Annealing Conditions			$\rho_C/10^{-5}$ ($\Omega \cdot cm^2$)	surface treatment, deposition method, comments	Ref.
				T (°C)	t (min)	ambient			
	200 °C	type substr							
WSi$_2$	100 sptr	1000; epi	10	1000	20	H$_2$	24	PECVD SiO$_2$/BOE	[106]

* In multilayer contacts, the metal on the left hand side was deposited first.
Abbreviations: n/r: not reported; e-beam: e-beam evaporation; therm.: thermal evaporation; sptr.: magnetron spattering; Ti/Al: layered deposition; Ti-Al: co-deposition from two sources; TiAl – alloy sputtering; epi: epitaxial layer; impl: implanted; vac: vacuum; UHV: ultrahigh vacuum (~3×10^{-10} Torr).

Fig. 16 shows reported specific resistances of ohmic contacts to *n*-type 6*H*- and 4*H*-SiC depending on SiC layer doping levels. Ohmic contacts are typically formed by the deposition of transition metals (possibly in combination with other metals, silicon or carbon) onto heavily doped silicon carbide (> 10^{18} cm^{-3}). The contact to 4*H*-SiC with $N_D > 10^{20}$ cm^{-3} demonstrate ohmic behavior as deposited while contacts to SiC layers with lower doping require high-temperature (> 950 °C) post-deposition annealing (PDA).

Note that the ρ_C values below ~2.5×10^{-6} $\Omega \cdot cm^2$ (shown by the horizontal solid lines in Fig. 15 and 16) should be considered as a rough estimation according to the calculations made in Section 3.4.

4.1 Nickel based ohmic contacts to *n*-type SiC

Although various transition metals have been studied extensively, nickel is the most widely used metal for fabrication of ohmic contacts to *n*-type SiC [109], [57]. Several groups have been successful in demonstrating nickel containing ohmic contacts to 4*H*-SiC with resistivities of around 1×10^{-6} $\Omega \cdot cm^2$ (see Fig. 17a).

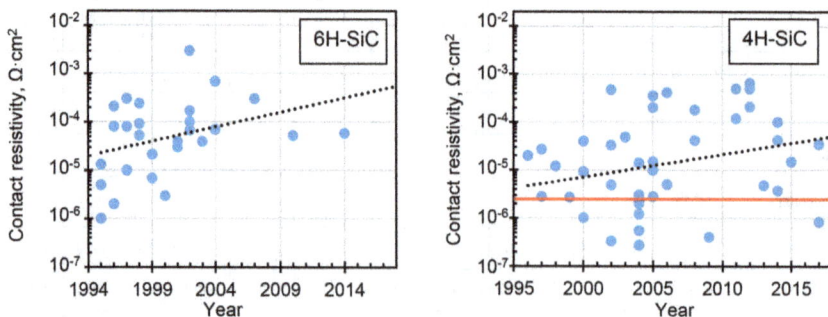

Figure 15.   Reported specific resistances of ohmic contacts to n-type SiC depending on a publication date. The exponential trendlines are shown by dotted lines.

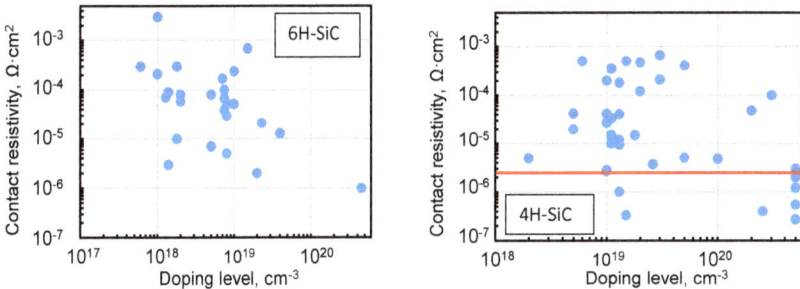

Figure 16. Reported specific resistances of ohmic contacts to n-type SiC depending on SiC layer doping level (data taken from Table 4 and Table 5).

There is an excess number of data points at doping level $\sim10^{19}$ cm^{-3} which is the maximum doping level of commercially available 4*H*-SiC epitaxial layers but, on average, nickel based ohmic contacts were fabricated to all layers with $N_D > 2\times10^{18}$ cm^{-3}. The resistivity of contacts with $\rho_C < 10^{-6}$ $\Omega\cdot$cm^2 were measured by TLM structures formed either on 800 nm thick epitaxial layer [62] or on layers with unreported thickness [63] and should be considered as a very rough estimation. Excluding these two data points, the lowest reported contact resistivity remained at the level of about 5×10^{-6} $\Omega\cdot$cm^2 at all doping levels above 2×10^{18} cm^{-3}.

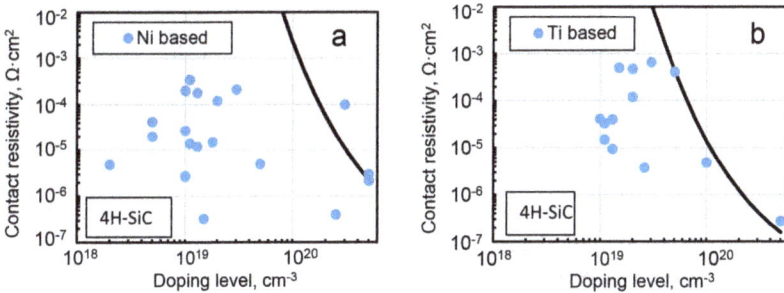

Figure 17. Reported specific resistances of (a) nickel and (b) titanium containing ohmic contacts to n-type 4H-SiC depending on SiC layer doping level (data taken from Table 4). Solid lines show ρ_C estimations by Eq. 13.

The solid line in Fig. 17a shows the ρ_C value for Ni contacts calculated by Eq. 13 taking into account Fermi level pinning described by Eq. 3a and Schottky barrier image-force lowering described by Eq. 4 (resulted in $\phi_{Bn} = 1.34$ eV at low doping and 0.89 eV at the solubility limit). It is clearly seen that the resistivites of Ni contact to 4H-SiC with doping levels $N_D > 10^{20}$ cm^{-3} (reported in [59] and [60]) are well matched by the calculations while at lower doping levels the contact resistivity is much lower than the ρ_C values estimated by Eq. 13 and almost independent on doping level. This may be attributed to the contacts with negative ϕ_{Bn} formed by energy band alignment but either nickel or products of its chemical interaction with SiC do not have work functions below 3.2 eV (χ_S value for 4H-SiC) required by Eq. 3a to form these ohmic contacts. This leads to the conclusion that the main reason for the conversion of Schottky to ohmic contacts during post-deposition annealing (which is a compulsory processing step at $N_D < 10^{20}$ cm^{-3}) is the change of electrical properties of the underlying SiC rather than being attributable to the electrical properties of resulted contact metallisation. However, the composition and microstructure of formed contact layers are important indicators of corresponding chemical reactions and interdiffusion processes during the PDA step which led to the required changes in the SiC layer and formation of ohmic contacts.

The first one of these indicators is formation of nickel silicide. It is well established that nickel silicides are formed during PDA as a result of SiC dissociation and chemical reaction between Ni and Si [110]. There are several thermally stable nickel silicides which may be created and some of them are $Ni_{31}Si_{12}$, Ni_2Si; NiSi and $NiSi_2$ The process starts at temperatures of about 600 °C with the creation of nickel-rich silicides ($Ni_{31}Si_{12}$) followed by formation of less metal-rich silicides at higher temperatures [111]. Nickel silicide contacts (mainly consisting of Ni_2Si phase) remain rectifying with Schottky barrier about 1.6 eV [112] when formed at temperatures below 950 °C. Contact behavior transfers from rectifying to ohmic when PDA is performed at higher temperatures, and only the Ni_2Si phase was found to be present in that contacts [111]. Fig. 18 shows the x-ray diffraction pattern of a Ni_2Si ohmic contact on 4H-SiC substrate. The 2θ scan with a fixed grazing incidence angle was taken by Bruker D8 Advance diffractometer equipped with a parabolic Göbel mirror and a conventional line focus tube with Cu anode (40 kV/40 mA). The phases detected by this x-ray phase analysis are Ni_2Si and C with traces of $NiSi_2$. To form this contact, a 5 nm thick Ti adhesion layer and 80 nm thick Ni layer were deposited subsequently on 4H-SiC substrate by the e-beam evaporation and annealed in vacuum at 1050 °C for 3 min. Nickel films can be deposited also by magnetron sputtering and thermal evaporation.

All evaporated Ni films are very strained and have a bad adhesion to the silicon carbide. They require employing either chromium or titanium adhesion layers with thickness of about 10 nm to deposit Ni films up to 100 nm thick. Magnetron sputtered nickel films are less strained and up to 300 nm thick nickel can be deposited directly on SiC substrate without any adhesion layer. In contrast to chromium which does not interact with SiC and remains spread in the reaction zone after annealing, titanium starts to interact with SiC at temperatures about 450 °C and forms very efficient diffusion barrier for nickel [110]. This leads to formation of less metal-rich silicide, $NiSi_2$, at the first moments of PDA due to limited nickel delivery to the front of solid-state chemical reaction zone [113] and this nickel disilicide is detected by the phase analysis shown in Fig. 18. The TiC layer remains stable during PDA marking the plane of the SiC surface in the initial structure before annealing (Kirkendall plane). In the case if all nickel is consumed during PDA, the TiC layer can be detected on the top of final contact structure by Auger profiling as it is shown in Fig. 19 [114]. Than is a direct proof of the fact that nickel is the most mobile species of the intermetallic compounds formed in the reaction zone [110] and chemical reaction between Ni and Si takes place at the interface between Ni_2Si and SiC. In other words, nickel diffuses to SiC through the reaction zone but not silicon to nickel.

The composite structure, consisting of the Ni_2Si as a low resistivity ohmic contact, and of the TiC as a diffusion barrier, promises a high-temperature stability of formed contacts but too thick titanium layer (~50 nm) deposited either before or after Ni results in a thick TiC layer formed in the interface between Ni_2Si and SiC [114]. This TiC layer significantly deteriorates electrical characteristics of formed ohmic contacts.

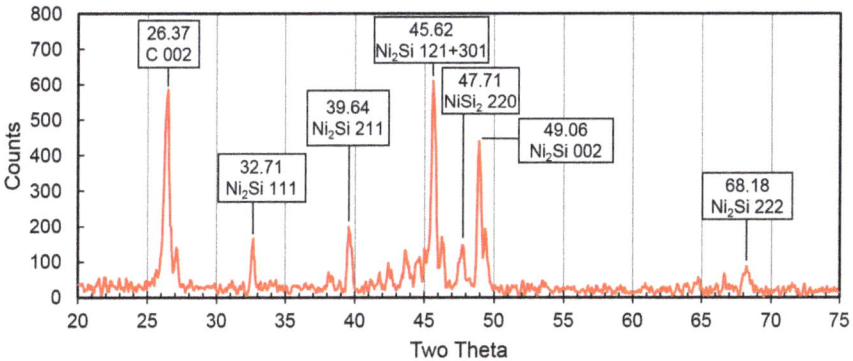

Figure 18. X-ray diffraction pattern of 80 nm thick Ni layer on 4H-SiC substrate annealed in vacuum at 1050 °C for 3 min. The 2θ scan with a fixed grazing incidence angle was taken by Bruker D8 Advance diffractometer equipped with a parabolic Göbel mirror and a conventional line focus tube with Cu anode (40 kV/40 mA).

The presence of the Ni_2Si phase at the interface with SiC is the first sign of ohmic contact formation. The Ni_2Si crystal phase has a work function of about 4.9 eV [30] and

Ni$_2$Si/4H-SiC contacts have to have a Schottky barrier height of 1.48 eV according to Eq. 3a. Electrical characterizations of Ni$_2$Si Schottky contacts revealed barrier heights of 1.3 eV for low doped 6H-SiC epitaxial layers [100] and ϕ_{Bn}=1.2 eV for low doped 4H-SiC [113]. Therefore, the presence of the Ni$_2$Si phase on its own cannot be the only factor in forming ohmic contacts on n-type SiC. As noted above, it indicates that the solid-state chemical reaction required for the formation of ohmic contacts took place during post-deposition annealing.

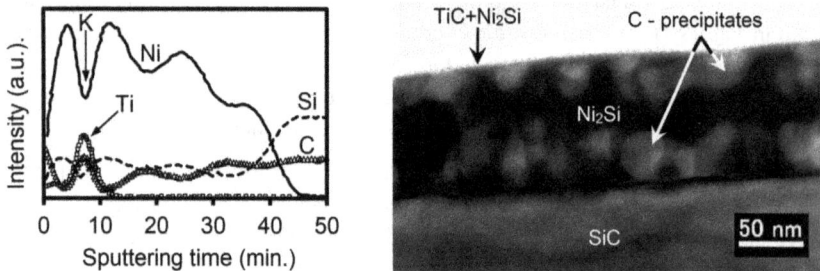

Figure 19. Auger profiles and TEM cross-section image taken from the Ni(150 nm)/Ti(4 nm)/SiC structure after annealing at 1040 °C for 800 s. Reprinted with permission from ref. [114]. Copyright (2006) IOP Publishing.

Another product of the solid-state chemical reaction between SiC and Ni is carbon. It is the most immovable compound in the reaction zone [115]. Its solubility in nickel and nickel silicide is very low: about 0.2% [116] and 0.8-1% [117], respectively. Therefore, carbon atoms remain in the place where they were released during dissociation of silicon carbide and agglomerate in precipitates clearly seen in the transmission electron microscopy (TEM) image shown in Fig. 19. These precipitates have a graphite structure detectable by x-ray phase analysis as it is shown in Fig. 18. Furthermore, the precipitates coalescence in a solid graphene-like film on the contact surface due to consumption of nickel during PDA [118], [119] and a few layer graphene film can be formed at the interface between contact and silicon carbide during the cooling stage of PDA [120].

Suggestions about the role of carbon in ohmic contact formation are ranging from carbon precipitates being responsible for contact degradation [68], [72] to carbon being a key factor in ohmic contact formation [92]. Both kinds of experiments (1) with deposition of pure Ni$_2$Si, [121] and Ni/Si [72] to avoid carbon in the contact and (2) with deposition of a 2.5 nm carbon film under the nickel to enrich the contact with carbon [122] were performed. They all revealed that post-deposition annealing is required to form ohmic contacts and no one of them demonstrated any significant reduction of contact resistivity.

As with Ni_2Si, it may be concluded that the presence of carbon either in the body of ohmic contact or at the interface between the contact and SiC cannot be the only factor in forming an ohmic contact but indicates that the solid-state chemical reaction required for the formation of ohmic contacts took place during post-deposition annealing.

An obvious verification of this suggestion was performed by Nikitina *et al.* [57]. They fabricated ohmic contacts on the C-face of a commercial *n*-type 4*H*-SiC wafer with a specific resistance of 0.023 $\Omega \cdot$cm. Before sample processing, the thin surface layer damaged by polishing was removed by reactive ion etching (RIE) to a depth of 10 μm, followed by sacrificial oxidation in wet oxygen at 1100 °C for 4 hours and subsequent oxide removal in HF. Titanium (4 nm thick) and nickel (170 nm thick) were deposited consecutively by thermal evaporation, patterned by the lift-off procedure and annealed in vacuum at 1040 °C for 800 sec. The contacts demonstrated an ohmic behaviour and were characterised by x-ray diffraction and Rahman spectroscopy to confirm that they contained Ni_2Si and carbon precipitates. After characterisation, the contacts were removed by etching Ni_2Si in HF:HNO_3 (1:3) and carbon in oxygen plasma. Then they were replaced by titanium "secondary" contacts patterned by the lift-off procedure and aligned with footprints remained in the SiC substrate after removal of the primary contacts. These "secondary" contacts demonstrated an as-deposited contact resistance as low as the original annealed contacts. This was a direct proof that the ohmic behaviour of nickel based contacts to SiC was defined by some changes in a thin SiC layer underneath the contact rather than by the composition of the contact body itself. It was supposed that positively charged carbon vacancies may be created in the near-interface region of the SiC resulting in a reduction of the Schottky barrier width and increasing the electron tunnelling probability. Existence of these carbon vacancies was not proven experimentally by other material characterization methods and the nature of changes in the thin SiC layer underneath the contact still remains unclear. Nevertheless, the "secondary" ohmic contacts were formed also on both 6*H*- and 4*H*-SiC polytypes, on Si- and C-faces using Au, Pt, Ta, or Ni as a second metal [123], [124]. These contacts remained stable when subjected to thermal stress at 300 °C in air for 1400 hours [125].

It is commonly assumed that nickel based ohmic contacts have two main drawbacks: (1) high roughness of the contact-silicon carbide interface and (2) Kirkendall voids and graphite precipitates in the contact body. The interface roughness was measured in [57] after nickel silicide contacts removal. The contacts were formed by thermal deposition on 4 nm thick Ti and 170 nm thick nickel by thermal evaporation followed by PDA in vacuum at 1040 °C for 3 minutes. It was found that SiC was consumed for the depth of

70 nm and the roughness of the SiC surface was around 10 nm root-mean square (rms), which is much higher than that measured in between the contacts pads (0.5 nm). This high roughness could contribute to decreasing contact resistivity but the effect of the interface roughness on contact resistivity and reliability was not studied directly.

To mitigate the second problems, multilayer Ni/Si (150 nm) films with Ni/Si ratio corresponding to Ni_2Si composition were deposited on $6H$-SiC [68]. These contact still required the PDA (950/10 min in Ar) to demonstrate ohmic behavior. It was supposed that the silicon additive should reduce silicon consumption from SiC and release of carbon atoms during PDA. The resulted contacts had resistivity about 10 times higher than conventional contact fabricated by annealing Ni layer on SiC. Decreasing of the number of Kirkendall voids was observed in TEM images but ageing tests for 100 hours at 500 °C in nitrogen did not show noticeable reliability improvement. Nakamura *et al.* [126] deposited Si (92.5 nm) and Ni (25 nm) films to produce the stoichiometric $NiSi_2$ alloy. The contacts were annealed at 950 °C for 10 min in Ar to get ohmic behavior and demonstrated resistivity comparable with literature data for contacts formed by nickel annealing. The absence of Kirkendall voids and graphite precipitates was confirmed by TEM but the contacts reliability was not tested.

Recently, Ervin *et al.* [124] investigated the effect of tungsten addition to nickel contacts. It was supposed that tungsten should reduce Kirkendall voids and graphite precipitates by formation of WC. The metal films were sputtered on Si face $4H$-SiC wafer ($N_D=1\times10^{18}$ cm^{-3}) and annealed at 1000 °C for 2 min in N_2 ambient. The SiC/Ni(54 nm)/W(30 nm) and SiC/Ni(74 nm) structures were processed side by side. Although the resistivity measurements by CTLM are not correct on a bare wafer, the direct comparison on the same wafer shown 10 times lower resistivity of Ni/W contacts than of Ni counterparts. The contacts were evaluated by XRD, XPS, TEM and AFM. It was found that the resulted contact structure is bilayer $SiC/WC/Ni_2Si$ with no voids, no carbon inclusions and better surface and interface roughness than in conventional SiC/Ni_2Si contacts. 300 nm thick Pt and 200 nm thick Au were sputtered on SiC/Ni and SiC/Ni/W for wire bonding and aging tests. The SiC/Ni/W contacts demonstrated good bond pull test results and ageing tests and significantly better long term thermal stability than SiC/Ni contacts. They remained relatively stable in atmospheric environment at 300 °C during 700 hours demonstrating resistivity linear increasing with ageing time while SiC/Ni degraded abruptly after 200 hours.

4.2   Practical tips and process compatibility of nickel silicide ohmic contacts

The depth of silicon carbide consumption during PDA can be calculated by Eq. 30 and it is roughly equal to the thickness of a deposited nickel layer in the case when all nickel has reacted. This SiC consumption has to be taken into account in a correct device design. The magnetron sputtered nickel is consumed during PDA with the rate of about 0.3 nm/s at 1050 °C (measured for RTP in vacuum for 10 min of 270 nm thick Ni film on SiC) and this value can be used for the first estimation of a required annealing time.

In case if the opposite site of SiC wafer is polished and not covered, the contact/SiC interface can be observed through the wafer. The blackening of nickel metallisation at the contact/SiC interface after PDA is the first sign of contact conversion to ohmic.

A mixture of sulfuric acid and hydrogen peroxide (volume ratio 4:1) can be used to selectively remove the unreacted Ni from nickel silicide [127] if it is required by device processing.

As it was mentioned in the previous section, a solid graphene-like carbon film is formed on the top of Ni_2Si contact or nickel if it has not reacted completely. This film makes an adhesion of any metal to formed ohmic contacts very poor. It has to be removed if subsequent metal deposition is required for further wire bonding, interconnection or probe testing. Sintering in H_2 for 20 min at 450 °C followed by slight etching using a BHF solution was used in [63] to remove this carbon film. It also can be sputtered by RIE in argon plasma (14 sccm, 20 mTorr, 300 W, 3 min).

Using nickel-based metallization to form ohmic contact may results in a significant simplification of SiC device fabrication as nickel can be used as a mask for RIE of SiC in a fluorine containing plasma. The selectivity of the e-beam evaporated nickel mask to SiC is about 70 (RIE in SF_6/O_2 = 52 sccm / 18 sccm, 70 mTorr, 300 W, 410 V dc). The magnetron sputtered nickel is etched much faster by RIE and has to be sintered in forming gas at 500 °C for few minutes before using as a mask. The Ni_2Si formed by PDA has very low selectivity to SiC and cannot be used as a mask for SiC RIE.

4.3   Nickel free ohmic contacts to *n*-type SiC

Many different metals and nickel free compound films were tested as contacts to *n*-type. There are several reasons for these studies. The first one is a search for materials with lower work functions than one of nickel and nickel silicides. They should form contacts with lower resistivity according to Eq. 13. These materials include, for example, titanium, niobium, hafnium, tantalum. The second objective is to find materials which can form contacts with better thermal and mechanical properties than nickel based contacts by reducing formation of Kirkendall voids and graphite precipitates. A material of this kind could be, for example, rhenium which remains stable in contact with silicon carbide at temperatures up to SiC sublimation temperatures [105] or titanium which forms both silicides and carbides when interacting with silicon carbide. The third motivation for these studies is to find metals or combination of metals which can form ohmic contacts simultaneously to both n- and p-type SiC. This may greatly reduce process complexity of some specific devices like SiC trenched and implanted vertical-channel junction field-effect transistors (TI-VJFET) [128]. Titanium is a component of most widely used ohmic contacts to p-type SiC and it is worth to test it as a material for formation of ohmic contacts to n-type SiC.

Apparently, titanium deserves consideration for any of these reasons. That is why the vast majority of publications about nickel free metallisation schemes are about titanium based ohmic contacts to n-type SiC. Table 4 and Table 5 list selected published data for titanium based ohmic contacts. Fig. 17b shows reported specific resistances of titanium based ohmic contacts to n-type $4H$-SiC depending on SiC layer doping levels. Similar to nickel based contacts (shown in Fig. 17a), the solid line in Fig. 17b shows the ρ_C value for Ti contacts calculated by Eq. 13 taking into account Fermi level pinning described by Eq. 3a and Schottky barrier image-force lowering described by Eq. 4 (resulted in ϕ_{Bn}= 0.92 eV at low doping and 0.51 eV at the solubility limit). It is clearly seen that the resistivites of Ti contact to $4H$-SiC with doping levels $N_D > 5 \times 10^{19}$ cm^{-3} are well matched by the calculations while at lower doping levels the contact resistivity is much lower than the theoretical ρ_C values estimated by Eq. 13 and almost independent on doping levels. The titanium work function is lower than the one of nickel (4.1 eV vs. 4.71 eV) but not low enough to form contacts with negative ϕ_{Bn}. As it was done for nickel based contacts, it may be assumed that the chemical reaction between Ti and SiC results in some changes of thin silicon carbide layer under the interface providing additional positive charge effectively reducing the width of the Schottky barrier. In average, specific resistance of titanium based ohmic contacts to n-type $4H$-SiC is noticeably higher than of nickel based contacts as can be seen from Fig. 17.

Titanium starts to interact with SiC at temperatures about 450 °C forming TiC. Released silicon diffuses through the reaction zone and forms Ti_5Si_3. This bi-layer structure was observed in contacts formed by PDA of Ti films at temperatures up to 950 °C [100]. Only the ternary phase Ti_3SiC_2 was observed in contacts annealed at higher temperatures. La Via *et al.* [100] compared resistivities of SiC/TiC/Ti_5Si_3 and SiC/Ti_3SiC_2 as well as test SiC/Ni_2Si contacts fabricated side by side on the same $6H$-SiC wafer (N_D=7.4×10^{18} cm^{-3}) using test structures with the same patters. They found that the TiC/Ti_5Si_3 bilayer contacts had specific resistance higher than Ti_3SiC_2 contacts (1×10^{-4} and 6.7×10^{-5} Ω·cm^2, respectively) and both contacts had higher resistivities than Ni_2Si (3.6×10^{-5} Ω·cm^2). It is worth to note that the Ti_3SiC_2 contacts were formed by PDA in vacuum at 1000 °C for 2 hours. This process requires a much higher thermal budget than is required for Ni_2Si contacts because titanium carbide being a very efficient diffusion barrier [110].

Buchholt *et al.* [82] grew Ti_3SiC_2 layers epitaxially on both *n*- and *p*-type 4*H*-SiC using magnetron sputtering from three separate targets. Depositions were carried out at 800 °C on Si face of 4*H*-SiC wafer with 0.8 μm thick *n*-type epilayer doped to 1.9×10^{19} cm^{-3} level. As-deposited films showed no ohmic behavior and required RTP at 950 °C for 1 min in Ar to get the contacts with resistivities of 5×10^{-4} $\Omega \cdot$cm^2 which is close to the value reported by La Via *et al.* [100]. This confirms that alterations in the SiC surface due to the annealing process are primarily responsible for the ohmic behavior. It should be noted that the grown Ti_3SiC_2 layers did not form ohmic contacts to *p*-type SiC even after annealing.

Another product of solid-state chemical reaction between Ti and SiC is titanium carbide. Lee *et al.* [58] grew 150 nm thick epitaxial TiC on *n*-type 4*H*-SiC epitaxial layers (x_S=1 μm; N_D=1.3×10^{19}cm^{-3}) by simultaneous evaporation of Ti and C_{60} in UHV (3×10^{-10} Torr) at 500 °C. The motivation for this study could be the low work function of TiC with some surface orientations (3.7 eV for TiC (110) [41]) but TiC layers grown in this experiment had (111) orientation with the highest work function, $q\phi_M$=4.7 eV (see Table 2). Nevertheless, they demonstrated ohmic behavior as deposited. To measure contact resistivity by TLM, SiC substrates were patterned before TiC growth by RIE and TiC film was etched after deposition (in a mixture of NH_3:H_2O_2=1:5 with the etch rate of 30 nm/min). As-deposited TiC contacts had resistivity of 0.93×10^{-5} $\Omega \cdot$cm^2 which increased to 4×10^{-5} $\Omega \cdot$cm^2 after PDA at 950 °C for 3 min in 10%H_2/Ar. The first value is 10 times lower than that one reported by La Via *et al.* [100] for 6*H*-SiC/TiC/Ti_5Si_3 bilayer contacts formed by PDA at 900 °C of titanium films deposited by e-beam evaporation while the second one is of the same scale. It should be noted that the same TiC film formed ohmic contacts to *p*-type 4*H*-SiC epi-layer (x_S=1 μm; N_A> 1×10^{20} cm^{-3}) with ρ_C=1.1×10^{-4} $\Omega \cdot$cm^2 as deposited and 5.6×10^{-5} $\Omega \cdot$cm^2 after PDA at 950 °C for 3 min in 10%H_2/Ar ambient. This metallization scheme could be chosen for simultaneous formation of ohmic contacts to *n*- and *p*-type SiC although it requires very specific processing. Indeed, contact films deposition in UHV at high temperature on patterned substrate and using direct chemical etching instead of the lift-off procedure may significantly limit its application in real devices.

There is a vast variety of other metals and compounds free of nickel and titanium which were tested as an ohmic contacts to *n*-type SiC. The commonly provided justification of such studies is that the proposed metal or metal compound barely interacts with silicon carbide ensuring dense contact body and smooth high quality interface with SiC but should demonstrate ohmic behaviour due to its unique electrical properties. Two examples are described below and more can be found in references provided in Table 4 and 5.

Jang *et al.* [108] investigated tantalum carbide (TaC) as an ohmic contact to *n*-type SiC. It is thermodynamic compatibility with SiC (the ternary phase diagram Ta-Si-C predicts TaC to be in equilibrium with SiC at temperatures up to 1000 °C) and has a relatively low work function ($q\phi_M$=4.22 eV [36]). A 100 nm thick TaC film was sputtered at 200 °C on *n*-type epi-layer (0.5 μm; 8×10^{18}cm^{-3}) grown in Si-face of *n*-type 6*H*-SiC substrate. The PDA at 1000 °C for 15 min in vacuum was required to get ohmic contacts with ρ_C=3×10^{-5} Ω·cm^2. Although the results of TLM measurements on *n*-layer grown on *n*-substrate can be taken only as a rough estimation, this ρ_C value is comparable with other published results. The PDA thermal budged is significantly higher than the one for nickel based contacts. It should be noted that these contacts demonstrated very good thermal stability when tested with W/WC (100 nm/50 nm) protection layer. They showed no degradation after ageing in vacuum at 600 °C for 1000 hours and at 1000 °C for 500 hours.

Oder *et al.* [88] investigated niobium as an alternative transition metal to form ohmic contacts to silicon carbide. It is corrosion resistant and has a high melting point 2468 °C. In addition, niobium has the same low work function as titanium. The 200 nm thick contact layers were sputtered on 4*H*-SiC epitaxial layers (0.5 μm; 1.3×10^{19} cm^{-3}) and on 6*H*-SiC epitaxial layers (0.5 μm; 1.4×10^{18} cm^{-3}). The contacts were rectifying after deposition and demonstrated an ohmic behaviour after PDA at 1100 °C for 10 min. The formation of NbC$_x$ and Nb$_5$Si$_3$C at the interface was suggested as responsible for the ohmic behaviour. An average specific contact resistance of 1×10^{-6} Ω·cm^2 (limited by the TLM accuracy) and 3×10^{-6} Ω·cm^2 were measured 4*H*- and 6*H*-SiC samples, respectively. These ρ_C values are comparable to those obtained by using the Ni metallisation scheme but the fabricated niobium contacts had problems with reproducibility from sample to sample and with oxygen incorporation (20 at.%) during sputter deposition.

4.4 Formation of ohmic contacts to implanted *n*-type SiC

The maximum doping concentration in commercially available *n*-type 6*H*- and 4*H*-SiC epitaxial layers is about 10^{19} cm^{-3}. Although ohmic contacts where demonstrated on epitaxial layers with this N_D level and even lower, there is a noticeable decrease of contact resistivity with N_D increasing as it is seen in Fig. 16. Higher SiC doping in under contact regions in device structures can be obtained by ion implantation followed by post-implantation annealing (PIA). Implantation services are widely available while PIA requires specialized facility. It is a very specific process due to high temperatures needed for recrystallization of SiC surface and complete activation of implanted dopants in silicon carbide [25]. In the case of nitrogen implantation in silicon carbide, optimized PIA has to be performed at temperature of 1700 °C in argon ambient for 10 minutes with a special SiC surface protection [129], [65] to avoid its deterioration and out-diffusion of implanted impurities.

Even properly protected during PIA, the implanted region may have a significantly lower doping concentration close to the SiC surface due to non-zero minimum ion energy as it is shown by profile 3 in Fig. 20. Although the top SiC layer is usually consumed during the following sacrificial oxidation and PDA process when the ohmic contacts are formed, this recession should be avoided since the nature of changes in the thin SiC layer underneath the contact remains unclear. The flat doping profile at the SiC surface can be obtained by ion implantation through a capping layer. Fig. 20 shows SIMS profiles of nitrogen implanted in 4H-SiC epi-wafer at room temperature and annealed at 1700 °C in argon ambient for 10 min under carbon capping layer. Profiles 1 and 2 were obtained by ion implantation through a 30 nm thick Al capping layer and do not have the recess at the SiC surface. Ohmic contacts where fabricated on these samples by the e-beam evaporation of 5 nm thick Ti and 80 nm thick Ni layers which were patterned by the lift-off procedure to measure contact resistance by TLM. The contacts formed on samples with doping profiles 2 and 3 required annealing in vacuum at 1000 °C for 3 min for the conversion from Schottky to ohmic contacts. Two times lower value of ρ_C measured in samples with profile 2 in comparison to samples with profile 3 may be attributed to the slightly higher doping level as well as to the absence of recess in doping profile at the SiC surface.

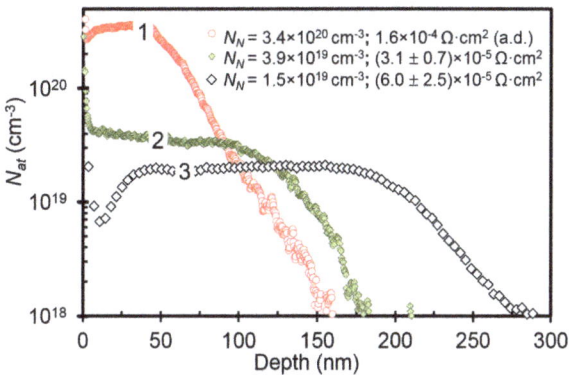

Figure 20.   SIMS profiles of nitrogen implanted in SiC and annealed at 1700 °C in argon ambient for 10 min. Profiles 1 and 2 were obtained by ion implantation through a 30 nm thick Al capping layer to avoid a surface recess in doping profile.

Advancing Silicon Carbide Electronics Technology I　　　　　　　Materials Research Forum LLC
Materials Research Foundations **37** (2018)　　　　　　doi: http://dx.doi.org/10.21741/9781945291852

The Ti/Ni contacts formed on the 4H-SiC layer with the nitrogen doping profile 1 in Fig. 20 demonstrated an ohmic behavior as-deposited. This leads to the conclusion that the doping level of 3.4×10^{20} cm^{-3} is high enough to have the current flow through this barrier defined by the electrons field emission and this is in a good agreement with the estimations made for Ti contacts to n-type 4H-SiC in Section 2 by Eq. 13. Indeed, Fig. 4 shows that the transition from TFE to FE dominating current mechanism has to take place at doping level of about 1.5×10^{20} cm^{-3}.

Formation of ohmic contacts to n-type 4H-SiC with even higher doping level was reported by Tanimoto *et al.* [59] and Na *et al.* [60]. The heavily doped 200 nm thick 4H-SiC layers with $N_D = 5 \times 10^{20}$ cm^{-3} were formed by phosphorus ion implantation at temperature of 500 °C through a 10 nm thick thermally grown oxide layer followed by PIA at 1700 °C in argon for 30 sec [130]. Several different metals were used to form contacts and all of them demonstrated an ohmic behavior as deposited. Fig. 21 shows resistivity of as-deposited contacts on n-type 4H-SiC layer with $N_D = 5 \times 10^{20}$ cm^{-3} formed by phosphorus ion implantation depending on metal work function. First, it should be noted that the resistivity of Ti contacts (2.7×10^{-7} $\Omega \cdot$cm^2) matches the estimation made in Section 2 for minimum available resistivity of titanium contacts to n-type 4H-SiC. Furthermore, the dependence of contact resistivity on metal work function follows the model based on the Schottky-Mott theory of metal-semiconductor contacts and described in Section 2. Indeed, substituting Eqs. 3a, 4 and 7 in Eq. 13 results in $\rho_{C(Ni)} / \rho_{C(Ti)} = 6.9$ while the experimental data shown in Fig. 21 give this ratio of about 11 which is a very good match to the estimations.

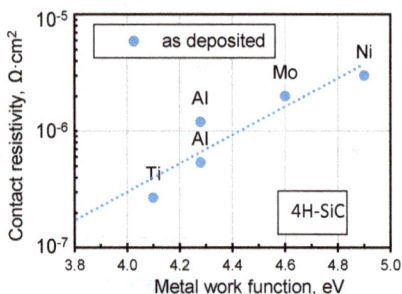

Figure 21. Resistivity of as-deposited contacts on n-type 4H-SiC layer with $N_D = 5 \times 10^{20}$ cm^{-3} formed by phosphorus ion implantation depending on metal work function.

Materials Research Forum LLC
doi: http://dx.doi.org/10.21741/9781945291852

Recently, Cheng *et al.* [131] demonstrated Ti/TiN/Al (100/10/300 nm) ohmic contacts to heavily doped 4*H*-SiC layers with $N_D=1\times10^{21}$ cm^{-3} which were formed by phosphorus ion implantation at room temperature. The SiC surface was prepared by *in-situ* etching in Ar inductively coupled plasma (ICP). The CTLM test structure patterns were made by RIE. The as-deposited contacts demonstrated an ohmic behavior with resistivity decreased to it minimum value $\rho_C=8.3\times10^{-7}$ Ω·cm^2 after PDA at 600 °C. This value of ρ_C is higher than the expected one for titanium according to the plot in Fig. 21 but it can be explained by lower SiC layer quality due to the implantation at room temperature.

5. Ohmic contacts to *p*-type SiC

The dominant mechanism of a charge carrier flow in Schottky barrier contacts to *p*-type 4*H*-SiC switches from TFE to FE at $N_A\sim1\times10^{21}$ cm^{-3} as it was estimated in Section 2. This value is about six times higher than the donor concentration required for changing from TFE to FE in SBC to *n*-type 4*H*-SiC. Nevertheless, contacts to *p*-type 4*H*-SiC with current mainly defined by FE still can be fabricated due to much higher aluminum solubility in SiC in comparison to phosphorus or nitrogen (about 2.0×10^{21} cm^{-3}, see Table 1). The minimum resistivity of ohmic contacts to *p*-type 4*H*-SiC is about 8×10^{-7} Ω·cm^2 as it was estimated in Section 2. This ρ_C value is 3 times higher than that one for ohmic contacts to *n*-type 4*H*-SiC. At the same time, the minimum resistivity of contacts to *p*-type 4*H*-SiC which can be correctly measured by TLM with decent accuracy is about 2.5×10^{-5} Ω·cm^2 as it was estimated in Section 3. It is 10 times higher than the one for ohmic contacts formed on *n*-type 4*H*-SiC due to higher bulk resistivity of *p*-type SiC.

Table 6 and Table 7 list ohmic contacts parameters, characterization and fabrication details reported for *p*-type 4*H*- and 6*H*-SiC, respectively. The data presented in these tables were taken from [8] and [9], critically revised, (some controversial data were excluded on the grounds of aforementioned estimations) and updated with recently published results. The contact resistivities listed in these tables were measured by TLM if another method is not indicated in the comments. It can be seen that the minimum contact resistivities published for all doping concentrations from $\sim10^{18}$ to $\sim10^{21}$ cm^{-3} are independent on N_A and are about 10^{-5} Ω·cm^2. Although there were several publications with reported contact resistivity $\sim10^{-6}$ Ω·cm^2, they are excluded from Table 6 and Table 7 for reasons explained in Section 3.

Table 6. Ohmic contacts to p-type 4H-SiC.

Metals*	Thickness (nm)	SiC layer info, x_S (nm)	N_A /10^{18} (cm^{-3})	Annealing Conditions T (°C)	t (min)	ambient	ρ_C /10^{-5} (Ω·cm^2)	surface treatment, deposition method, comments	Ref.
Al and Al/Ti based									
Al	160 e-beam	4000 epi	4.8	1000	2	vac	42		[132]
Al	500 therm.	220 Al+C impl at RT	600	950	5	Ar	6.1		[133]
Al/Ti	40/60	500 LPE	40	900	5	Ar	1.42	Ar+ ions	[134]
Al/Ti	117/30 sptr.	epi	10	1000	2	n/r	7	550C/10 min vac	[135]
Al/Ti	25/150; e-beam	200; epi	1	900	3	Vac	64		[136]
Ti/Al	50/190; e-beam/therm	5000; epi	4.5	1000	2	vac	2	sacr. ox. CTLM	[137]
Al/Ti/Al	160 in total (Ti 31wt.%) e-beam	4000; epi	4.8	1000	2	vac	33		[132]
Ti/Al	70/30 wt.% sptr/e-bam	1000; epi	40	900	5	1% H$_2$/Ar	1.42	Ar+ ions	[138]
Al/Ti/Ni	50/50/20 e-beam	x_s n/r; epi	4.5	800	30	UHV	220	sacr. ox. CTLM	[87]
Ti/Al/Ti/Pd/Ni	3/50/100/10/50 e-beam	300 LTLPE	150	1150	1	H$_2$	10	CTLM	[139]
Al/Ti/Pt/Ni	50/100/25/50 e-beam	1000 epi	7	1000	2	vac	9		[140]
Al/Ti/Pt/Ni	50/100/10/50 e-beam	200 epi	20	1000	2	vac	15	sacr. ox.	[65]
AlTi	x_m n/r; (Ti 30 wt.%) sptr	x_S n/r; Al impl	1000	1000	2	vac	20		[76]
AlSiTi	Si 2%; Ti 0.15%	epi	40	950	5	Ar	9.6	Ar+ ions	[141]
AlTi	200 (Ti 30 wt.%)	x_S n/r; epi	13	1000	2	vac	3	Polytype n/r	[142]
Ti/Al	160 e-beam; Ti - 31 wt.%	4000 epi	4.8	1000	2	vac	25		[132]
Ti50Al50	200 sptr.	1300 epi	10-40	1000	10	Ar	11		[43]
Ti/Al	25/95	2500+500 epi	100	850	1	Ar	4.8		[143]
Ti/Al	100/300	170 Al impl	74	950	1	Ar	14.5		[144]
Ti/Al	50/140	500 epi	20	900	5	N$_2$	6.6		[145]
Ni/Ti/Al	35/50/300; e-beam/therm	5000 epi	4.5	800	30	vac	7	sacr. ox.; CTLM	[137]
Ti/Al/Ni	70/200/50 sptr;	300 Al impl 600 °C	100	950	1	Ar	23	epitaxial TiC(111)	[146]
Ti/Al	300 total Ti (13-23 wt.%) e-beam	5000 epi	10	500+1000	30+2	UHV	1	sacr.ox, CTLM	[147]
Ti/Al/Si	20/30/30	x_S n/r; impl	24	1020	2.5	Ar	17	sacr. ox.	[79]
Ti/Al/W	70/200/50 e-beam	300 Al impl	100	1100	n/r	Ar	58		[148]
Other materials									
Al/Ni	6 wt%/50; e-beam	200; Al-impl	700	1000	2	Ar	52		[85]
Al/Ni	6/50; therm	epi, x_S n/r	7.2	1000	5	UHV	1200	CTLM	[71]
Al/Ni	15/50 sptr.	300 Al impl in 1000 p-epi	1000	850	1	Ar	1	CTLM	[149]
Al/Pd	30/70	500; LPE	40	900	5	Ar	4.08	Ar+ ions	[134]

Metals*	Thickness	SiC layer info, x_S	$N_A/10^{18}$	Annealing Conditions			$\rho_C/10^{-5}$	surface treatment, deposition method, comments	Ref.
				T	t	ambient			
(nm)	(nm)	(nm)	(cm⁻³)	(°C)	(min)		(Ω·cm²)		
Co/Al	10/40 e-beam/therm	5000; Epi	9	900	5	vac	40	CTLM	[150]
Ge/Ti/Al	10/15/75 at.% e-beam/therm	5000; Epi	4.5	600	30	vac	10.3	sacr. ox.	[151]
Ge/Ti/NiAl	24/32/144; sptr	300; Al impl on 3000 p-epi	100	600	30	vac	80		[152]
Ni	100; sptr	500 impl in 2500 p-epi	100	800	1	Ar	3		[153]
Ni	100; e-beam	x_S n/r; Al-impl at 400C	100	1000	1	N_2	100		[61]
Ni	100; sptr	300 impl in 1000 epi	1000	850	1	Ar	2	CTLM	[149]
NiV	x_m n/r; sptr. 7% V	x_S n/r; Al impl	1000	900	1	vac	8		[76]
Al/Ni	50/50	270 Al-impl	200	1000	2	Ar	700	sacr. ox.	[62]
Ni/Al	50/300 e-beam	5000 epi	6.4	1000	5-30	vac	9.5	sacr. ox.	[154]
Ni/Ti/Al	25/50/300 e-beam	5000 epi	3	800	5-30	vac	6.6	sacr. ox.	[154]
Pd/Ti	10/100 sptr	500; epi on 2500 p-epi	100	400 + 850	1.5 + 1	N_2	1.6	Four-point probe	[155]
Pd	100	500 LPE	40	900	5	Ar	4.82	Ar+ ions	[134]
Pd	80-150 sptr.	epi	50	15	5	N_2	35	Ar plasma	[156]
Pd/Ti/Pd	10/20/80 e-beam/sptr/e-beam	500; epi	40	900	n/r	1% H_2/Ar	2.9		[157]
Pt	100 sptr.	1000 epi	10	1100	5	n/r	15	550C/10 min vac	[135]
Si(B)/Pt	66/50 sptr.	1000 epi	10	1000	5	n/r	4.4	550C/10 min vac	[135]
Si/Pt	20/80 e-beam	200 epi	10	1100	3	Vac	58		[136]
Si/Al	80-150 sptr , thickn. ratio n/r	1000 epi	50	700	20	N_2	10	Ar plasma	[158]
Ti	140 at 300 C; in UHV	1000 epi	>100	0	0		34.4	UHV	[58]
Ti	140 sptr.at 300 C; in UHV	1000 epi	>100	950	3	10% H_2/Ar	77	UHV	[58]
Ti/Ni	10/20; e-beam	600; Al impl	100	950	1	N_2	130		[80]
Ti/NiAl	32/144; sptr	300; Al impl on 3000 p-epi	100	975	2	Ar	5.5		[152]
Ti-C	150 co-sptr.at 500 C; in UHV	1000 epi	>100	0	0		10.8	UHV; TiC epi	[58]
Ti-C	150 co-sptr.at 500 C; in UHV	1000 epi	>100	950	3	10% H_2/Ar	5.6	UHV; TiC epi	[58]
Ti-C	130 co-sptr.at 500 C; in UHV	300 Al-impl at 700C	20	500	n/r	10% H_2/Ar	2	UHV; TiC epi	[159]
W/Al	20/95	2500+500 epi	100	850	1	Ar	6.8		[143]
Ti/Si/Co	40/20/80 sptr	500 epi	3.9	850	1	vac	40	TLM without mesas	[160]

* In multilayer contacts, the metal on the left hand side was deposited first.
Abbreviations: n/r: not reported; e-beam: e-beam evaporation; therm.: thermal evaporation; sptr.: magnetron spattering; Ti/Al: layered deposition; Ti-Al: co-deposition from two sources; epi: epitaxial layer; impl: implanted; vac: vacuum; UHV: ultrahigh vacuum ($\sim 3 \times 10^{-10}$ Torr).

Fig. 22 shows reported specific resistances of ohmic contacts to p-type $6H$- and $4H$-SiC depending on SiC layer doping levels. As it was shown in Section 2, the lowest resistivity have to be demonstrated by ohmic contacts formed by metals with highest work functions. The estimations made by Eq. 13 for Pt (ϕ_M=5.7 eV) contacts to p-type $4H$-SiC (ϕ_{Bp}=1.19 eV at low doping and 0.56 eV at the solubility limit) are shown by the solid line in Fig. 22.

It is clearly seen that the contacts formed on $4H$-SiC with $N_A<5\times10^{20}$ cm^{-3} have resistivities noticeable lower than the estimated minimum ρ_C values and are almost independent on doping level. As in the case of ohmic contacts to n-type SiC it may be concluded that the resistivity of ohmic contacts to p-type SiC is defined mainly by alternating the properties of SiC beneath the contact during deposition or PDA rather than by the work function of a contact material. Indeed, the lowest specific resistances were obtained in the contacts to p-type SiC containing titanium, aluminum and products of their chemical interaction with silicon carbide and all these materials have work functions lower than platinum.

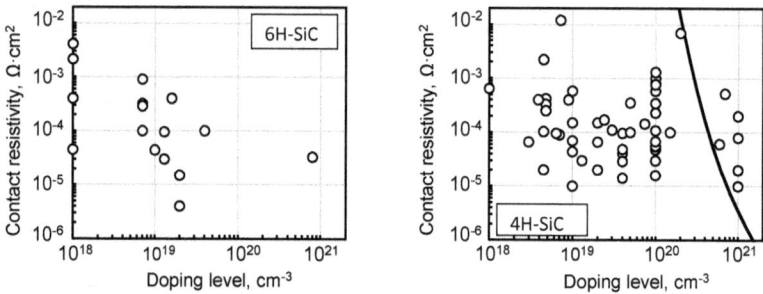

Figure 22. Reported specific resistances of ohmic contacts to p-type SiC depending on SiC layer doping level (data taken from Table 6 and Table 7).

Table 7. Ohmic contacts to p-type 6H-SiC.

Metals*	Thickness (nm)	SiC layer info, x_S (nm)	N_A/10^{18} (cm^{-3})	Annealing Conditions T (°C)	t (min)	ambient	ρ_C/10^{-5} ($\Omega\cdot$cm^2)	surface treatment, deposition method, comments	Ref.
Al	500 therm.	200 Al+C impl	800	950	5	Ar	3.3		[133]
Al/Ti	117/30 sptr.	epi	7	1000	2	n/r	10	550C/10 min vac	[135]
Al/Mo	60/60; sptr	substr	1	1200	0.67	N$_2$-H$_2$	4.5	Two-point probe	[161]
Al/Ta	60/160; sptr	substr	1	400+1000	3+0.33	Ar	40	Two-point probe	[161]
Al/Ti	150/30	Epi	16	900	4	N$_2$	40	epi Al$_3$Ti covered by Ti$_3$SiC$_2$	[162]

Metals*	Thickness (nm)	SiC layer info, x_S (nm)	N_A /10^{18} (cm^{-3})	Annealing Conditions			ρ_C /10^{-5} ($\Omega \cdot$cm^2)	surface treatment, deposition method, comments	Ref.
				T (°C)	t (min)	ambient			
AlTi	300-500 sptr Ti% n/r	1800 epi	20	1000	5	Ar	1.5	CTLM	[163]
Al1%Si/Ti	350/80 sptr	1400 Al-impl at 305 °C	40	1000	2	Ar	10		[164]
Co/Si	50/160 e-beam	1000 epi	20	500+900	300+1 20	Vac	0.4	$L_T \sim 0.9$ μm	[165]
CrB$_2$	200 sptr	1000 epi	13	1100	15	Vac	9.58		[166]
Mo	550 sptr	substr	1	1000	0.33	N$_2$-H$_2$	408	Two-point probe	[161]
Pt	100 sptr.	1000 epi	7	1100	5	n/r	90	550C/10 min vac	[135]
Si(B)/Pt	50/66 sptr	1000 epi	7	1000	5	n/r	32	550C/10 min vac Si deposited at 500C	[135]
Si(B)/Pt	50/66 sptr	500 epi	7	1100	5	Vac	28.9	550C/10 min vac Si deposited at 500C	[167]
Ta	160 sptr	substr	1	1100	10	N$_2$-H$_2$	213	Two-point probe	[161]
Ti	300	x_S n/r; epi	13	800	1	vac	3		[168]
TiN	100 PLD at 600 °C	x_S n/r; epi	10				4.4	focused ion-beam	[169]

* In multilayer contacts, the metal on the left hand side was deposited first.
Abbreviations: n/r: not reported; e-beam: e-beam evaporation; therm.: thermal evaporation; sptr.: magnetron spattering; PLD: pulsed laser deposition; Ti/Al: layered deposition; Ti-Al: co-deposition from two sources; TiAl – alloy sputtering; epi: epitaxial layer; impl: implanted; vac: vacuum; UHV: ultrahigh vacuum (~3×10^{-10} Torr).

5.1 Al and Al/Ti based contacts to *p*-type SiC

Aluminium, which is an acceptor impurity in SiC, was the first metal examined as a material for ohmic contacts to *p*-type SiC [170]. The Al-Si eutectic (11.3 at.% Si, melting point 577 °C) was alloyed in vacuum at temperature about 1000 °C in low doped hexagonal *p*-type SiC with very high bulk resistivity of 100 Ω·cm. The contacts demonstrated ohmic behavior but contact resistivity was not measured.

Decent ohmic contacts to *p*-type SiC can be fabricated by thermal evaporation of thick aluminium films (500 – 1500 nm) on silicon carbide substrate heated to 600 °C. Although, aluminium starts to interact with SiC at lower temperatures, about 500 °C, forming Al$_4$C$_3$, reaction rate was not high enough to consume all deposited aluminium. These contacts were ohmic as deposited, they had a good adhesion and uniform morphology. Their resistivity and structure were not investigated thoroughly, but this metallisation scheme was used widely at earlier days of SiC technology to fabricate 6*H*-SiC light emitting diodes and various test devices. This aluminium ohmic contacts demonstrated acceptable high-temperature operation and thermal stability when used, for example, for fabrication 6*H*-SiC dinistors operating at temperatures up to 500 °C [56] and 6*H*-SiC pulsed avalanche diodes with power density of 90 MW/cm^2 [171]. Due to metal deposition at very high temperature, this method is barely compatible with conventional device processing.

Aluminium ohmic contacts to p-type SiC which were formed by conventional processing with metal deposition at room temperature followed by PDA were probably first reported by W. von Muench *et al.* [172]. The contact to p-type 6H-SiC Lely crystals with N_A=2×10^{18} cm^{-3} were prepared by evaporation of aluminum/silicon (eutectic) and alloying at 1100 °C. *I-V* characteristics of p-n junction diodes were provided but no details of contact structural and electric characterization were reported. It should be noted that evaporation from an alloy produces a film of that alloy component which has higher vapor pressure which is aluminum in this particular case.

Johnson *et al.* [132] suggested that pure aluminium films form ohmic contacts to p-type SiC after PDA at temperatures about 1000 °C due to creation of Al$_4$C$_3$ phase at the interface. This is a compound semiconductor with a bandgap of about 3 eV [173]. It has to be heavily doped by silicon released during the interfacial reaction between aluminum and silicon carbide or by lattice vacancies resulting in creation of a purely tunneling contact. Unfortunately, aluminium contacts formed by PDA at temperatures of about 1000 °C suffer from significant loss of aluminium during annealing and formation of etch pits in silicon carbide. These pits are distributed non-uniformly causing problems with reproducibility of contact resistance from sample to sample. They can be about 100 nm deep resulting in the damaging of the underlying device structure. In addition, Al$_4$C$_3$ phase formed during PDA is known to be chemically unstable, decomposing into methane gas and aluminum hydroxide in such mild reagents as moist air at room temperature [132].

Further development of ohmic contacts to p-type SiC happened in the earlier ninetieth when AlTi alloy was proposed instead of pure aluminium [174]. An addition of about 10 wt.% of titanium increased the alloy melting point above 1000 °C making possible high temperature PDA without significant loss of metallisation as well as high temperature operation of fabricated contacts. The earliest data on AlTi contacts resistivity was published by Crofton *et al.* [163]. They sputtered 300-500 nm thick AlTi alloy (Ti content was not reported) and formed ohmic contacts on 6H-SiC epitaxial layers (x_S=1.8 μm; N_A=2×10^{19} cm^{-3}) with ρ_C=1.5×10^{-5} Ω·cm^2 after PDA at 1000 °C for 5 min in Ar.

Later, Crofton *et al.* [142] optimised the AlTi alloy content. The films with 10, 30, 40 and 81 wt.% of Ti were sputtered on epitaxial layers with N_A=1.3×10^{19} cm^{-3}. They found that the layer of sputtered AlTi alloy with 10% titanium by weight had to be at least 250 nm thick to produce ohmic contacts due to Al loss during PDA (1000 °C for 2 min in vacuum). The contacts still had bad morphology due to spikes into the semiconductor layer. The sputtered alloys with 40 and 81 wt.% of titanium did not produce ohmic contacts while the contacts formed by PDA of AlTi alloy with 30 wt.% of Ti demonstrated the lowest resistivity (ρ_C=5×10^{-5} Ω·cm^2), good morphology and reproducibility.

Abi-Tannous *et al.* [43] sputtered TiAl alloys with 31, 43 and 63 wt.% of Ti on *p*-type 4*H*-SiC layers (x_S=1.3 μm; N_A=1.5×10^{19} cm^{-3} and 3×10^{19} cm^{-3}). The contacts were annealed at temperatures from 900 °C to 1200 °C in argon ambient for 10 minutes. All of them demonstrated ohmic characteristics when annealed at temperatures 1000 °C and higher and had minimum resistivities (from 1.1 to 4.1×10^{-4} Ω·cm^2) after PDA at 1000 °C. Only the contacts made of the alloy with 63 wt.% of Ti were formed on epi-layers with highest doping concentration (3×10^{19} cm^{-3}) and probably for this reason demonstrated lowest resistivity (ρ_C=1.1×10^{-4} Ω·cm^2) which was reducing to 1×10^{-5} Ω·cm^2 when measured at 600 °C confirming the TFE current flow mechanism in these contacts.

Nakatsuka *et al.* [147] investigated the effect of Ti content on resistivity of contacts prepared by the e-beam evaporation of multi-layer Ti/Al films (up to six pairs of Ti/Al bilayers) on a 4*H*-SiC epitaxial layer (5 μm; N_A=1×10^{19} cm^{-3}). The 300 nm thick contact films had 13; 20; 23 and 25 wt.% of titanium. Contact annealing was performed in UHV in two steps: the first one at 500 °C for 30 min to form the AlTi alloy and then the second step at 1000 °C for 2 min to facilitate its interaction with silicon carbide. It was found that the contact resistivity does not depend on the number of layers but only on the titanium content in the whole deposited film. The contacts with 25 wt.% of titanium demonstrated non-ohmic behaviour and ρ_C ~1×10^{-5} Ω·cm^2 was measured when Ti content was below that value.

Based on all empirical data available for Ti/Al contacts to *p*-type SiC, it may be suggested that the most reproducible contacts with minimum resistivity are formed when titanium content in the initial film is about 30% by weight and PDA is performed at temperature of 1000 °C. In this case the Ti$_3$SiC$_2$ ternary compound is often observed in annealed layers and it is indicated as a key factor for the ohmic behaviour. Contacts annealed at lower temperatures (<950 °C) tend to form bilayer TiC/Ti$_5$Si$_3$ structures [100] and contacts annealed at higher temperatures (>1100 °C) tend to form bilayer TiC/Ti$_3$SiC$_2$ structures [43]. In terms of concentrations, samples with excess aluminum form Al$_4$C$_3$ and those with excess titanium form Ti$_5$Si$_3$ while Ti$_3$SiC$_2$ covered by Al$_3$Ti was observed in samples with 31 wt.% of Ti after annealing [132].

The Ti$_3$SiC$_2$ ternary compound is a material with an unusual combination of metallic properties such as high thermal and electrical conductivity with ceramic properties such as chemical inertness and thermal stability. It is a narrow-gap semiconductor with a theoretically determined indirect bandgap of 0.12 eV [175] and relatively high electron affinity of 5.07 eV [44] which is intermediate between Ti work function (4.1 eV) and the

Advancing Silicon Carbide Electronics Technology I   Materials Research Forum LLC
Materials Research Foundations 37 (2018)   doi: http://dx.doi.org/10.21741/9781945291852

value of ϕ_S for p-type $4H$-SiC (6.4 eV). An interfacial layer of Ti_3SiC_2 had to lower the barrier height enhancing the thermionic emission but pure Ti_3SiC_2 grown at 800 °C using magnetron sputtering from three separate targets did not form ohmic contacts to p-type $4H$-SiC even after post-growth annealing [82]. This suggests the critical role of aluminum in forming ohmic contacts when Ti_3SiC_2 is grown from TiAl alloys or Ti/Al multilayered films during PDA on a SiC substrate. The main function of Al was determined to form a liquid alloy that promotes the reaction between titanium and silicon carbide which results in formation of Ti_3SiC_2 at the SiC surface. This Ti_3SiC_2 layer was found to be strongly inhomogeneous and interrupted by Al-rich regions independently on being formed on an as-grown epitaxial SiC surface [176] or on a heavily implanted SiC layers (N_{Al}=7.4×10^{19} cm^{-3}) [144]. Gao et al. [44] observed reaction induced interfacial defect states in the near-interface SiC by depth-resolved cathodoluminescence. They were believed to further reduce the barrier height and cause the ohmic contact formation. Apparently, the combination of all these three factors (Ti_3SiC_2 grains, Al-rich inclusions and interfacial defect states in the near-interface SiC) are responsible for ohmic behavior of contacts formed by PDA of Al/Ti films (with about 30 wt.% of Ti).

Both titanium and aluminium tend to oxidise very easily during post deposition annealing even when the process is carried out in a high vacuum or high purity inert atmosphere. The Al/Ti contacts need to be protected by some oxidation resistant capping layer. Vivona et al. [148] used 50 nm thick tungsten layer on top of Ti/Al (70/200 nm thick). The contacts became ohmic after PDA at 1100 °C with resistivity of 5.8×10^{-4} $\Omega\cdot cm^2$ which is almost 10 times higher than the resistivity of Al/Ti formed on the same implanted $4H$-SiC layers (N_{Al}=10^{20} cm^{-3}) [144]. It was found that tungsten diffused towards the interface where formation of $W(SiAl)_2$ and TiC was observed instead of Ti_3SiC_2. Using 50 nm thick nickel instead of tungsten deposited on the same Al/Ti films resulted in ohmic contacts with lower resistivity, ρ_C=2.3×10^{-4} $\Omega\cdot cm^2$ after PDA at 950 °C for 1 minute in argon. The contact were very rough (50 nm rms) with agglomerates of Al_3Ni_2 alloy on top of Ti-C-Si layer and epitaxial TiC(111) at the interface with SiC.

Vassilevski et al. [140] employed platinum barrier layer to prevent the nickel diffusion into the Al/Ti contact layer. They deposited Ti/Al/Pt/Ni (50/100/25/50 nm thick) multi-layer structure by e-beam evaporation on a commercial p-type $4H$-SiC epitaxial layer with N_A=7×10^{18} cm^{-3}. Ohmic contacts with ρ_C=9×10^{-5} $\Omega\cdot cm^2$ were obtained after PDA at 1000 °C for 2 minute in vacuum. Thickness of platinum barrier was optimised to get the best morphology of the resulted contact layer. The absence of nickel diffusion into the Al/Ti contact layer was confirmed by AES.

The contacts based on aluminum-titanium alloy are presently the most widely used in SiC device structures. Fig. 23a shows selected published data on specific resistances of Ti/Al based ohmic contacts to p-type $4H$-SiC depending on a doping level. This metallization scheme provides ohmic contacts with resistivities of about 5×10^{-5} $\Omega \cdot cm^2$ to both epitaxial and implanted layers with doping concentrations exceeding 10^{18} cm^{-3}. This contact resistivity is acceptably low for fabrication SiC devices taking into account relatively high bulk resistivity of silicon carbide in comparison to silicon. Nevertheless, this ρ_C value is still about one order of magnitude higher than the minimum theoretical value estimated in Section 2. It is the main motivation for further search of alternative metallisation schemes and contact formation techniques.

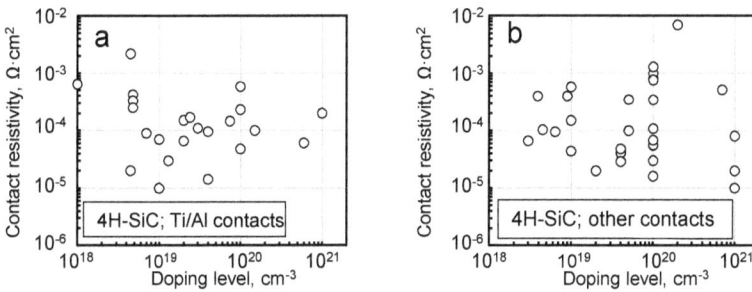

Figure 23. Reported specific resistances depending on 4H-SiC layer doping level (data taken from Table 6). Left: Al/Ti based ohmic contacts to p-type 4H-SiC. Right: contacts to p-type 4H-SiC formed by other materials.

5.2 Practical tips on fabrication Al and Al/Ti based contacts to p-type SiC

Consumption of silicon carbide due to its interfacial reaction with Al/Ti alloy during PDA has to be taken into account in a design of SiC devices with relatively thin p-type layers. In contrast to nickel contacts to n-type SiC, it cannot be easily calculated because of this reaction may have several different products, namely, Ti_3SiC_2, Ti_5Si_3, TiC. Furthermore, the amount of reacted SiC can be significantly lower due to aluminium evaporation during PDA. As a rough estimation, Ti/Al (50/100 nm thick) layer covered by Pt/Ni (25/50 nm thick) protection layer consumed 110 nm of SiC during PDA at 1000 °C for 2 minute in vacuum [140].

Pure aluminum can be used as a mask for RIE of SiC in a fluorine containing plasma. The selectivity of a e-beam evaporated aluminum mask to SiC is about 20 (RIE in SF_6/O_2 = 52 sccm / 18 sccm, 25 mTorr, 200 W, 455 V dc). The pure titanium, Ti_3SiC_2 and Ti_5Si_3

have very low selectivity to SiC. For this reason Al/Ti contacts after PDA cannot be used as a mask for SiC RIE.

Thick (>1 μm) aluminum thermally evaporated on SiC substrate at 600 °C can be used as a mask for SiC etching in molten KOH. To form mesa structures, the deposited Al film has to be patterned using the contact photolithography and etching in the aluminum etchant ($HNO_3/H_3PO_4/CH_3COOH/H_2O$). The SiC etch process has to be conducted at ~450 °C in KOH melt in a nickel crucible. The melt has to be preliminary dehydrated at least for two hours at this temperature. The Al mask had the etch selectivity to SiC exceeding 1:20 when used to pattern the C-face of SiC. This process was widely used in SiC device processing in the early days of SiC technology for formation of mesa-structures and for back-side etching.

Evaporated aluminum have a bad adhesion to the silicon carbide and deposition of multilayer Ti/Al structures on a SiC substrate has to always start with a titanium layer.

Aluminum expands during solidification. This causes a graphite (FABMATE) crucible used in e-beam evaporators to break apart just after few deposition runs. To prolong the crucible lifetime, aluminum can be evaporated from a Ti20Al80 wt.% alloy. A pure aluminum film will be deposited due to higher Al vapor pressure in this case.

5.3   Ohmic contacts to *p*-type SiC based on other metallization schemes

Various metals and compound films tested as alternative contacts to *p*-type SiC are listed in Table 6 and Table 7. Fig. 23b shows reported specific resistances depending on a doping level for these contacts (data taken from Table 6). Apparently, the minimum contact resistivities at the same doping levels are slightly lower for the alternative contacts when compared with Al/Ti based ones (shown in Fig. 23a) but still significantly higher than the minimum theoretical value estimated in Section 2 ($\sim 10^{-6}$ $\Omega \cdot cm^2$). Several representative examples are described below and more can be found in references provided in Table 6 and 7.

There are several motivations for the search of alternative metallisations for ohmic contacts to *p*-type SiC. First, metals with higher work functions (e. g. Pt, Pd, Co, Ni) have to form contacts with *p*-type SiC with lower barrier than aluminum and titanium do and, hence, have to have a lower contact resistance. But it should be noted that the Ti_3SiC_2 ternary compound, which is formed during optimized PDA of Ti/Al contacts to SiC, has itself a relatively high electron affinity of 5.07 eV [44] and only platinum has a higher work function (5.7 eV).

Sputtered 100 nm thick pure platinum was tested as a contact to *p*-type SiC [135]. Ohmic contacts with $\rho_C = 9 \times 10^{-4}$ $\Omega \cdot cm^2$ were obtained after PDA at 1100 °C for 5 minutes on a

$6H$-SiC epitaxial layer (1 μm; N_A=7×10^{18} cm^{-3}) and with ρ_C=1.5×10^{-4} Ω·cm^2 on a $4H$-SiC epitaxial layer (1 μm; N_A=1×10^{19} cm^{-3}). Both these values exceeded the resistivity of Al/Ti based contacts fabricated and characterised side by side on the same epi-wafers. Additionally, the contacts displayed a very rough morphology and inhomogeneous distribution of different platinum silicide phases through the contact area.

To solve this problem, the boron doped silicon (50 nm thick) was sputtered on SiC substrate at 500 °C before deposition of 66 nm thick platinum [135], [167]. The thickness ratio was chosen to be stoichiometric to the PtSi phase. This phase is thermodynamically stable with silicon carbide at temperatures exceeding 1000 °C, it has an excellent electrical conductivity and work function of 4.97 eV [177]. Ohmic contacts were obtained after PDA at 1000 °C for 5 minutes (at lower temperature than in the case of pure platinum). They demonstrated very smooth morphology and only a single platinum silicide phase was detected. The contact resistivity was measured ρ_C=3×10^{-4} Ω·cm^2 on a $6H$-SiC epitaxial layer (1 μm; N_A=7×10^{18} cm^{-3}) and ρ_C=4.4×10^{-5} Ω·cm^2 on a $4H$-SiC epitaxial layer (1 μm; N_A=1×10^{19} cm^{-3}). Both these values were comparable with the resistivity of Al/Ti based contacts fabricated and characterised side by side on the same epi-wafers (1×10^{-4} and 7×10^{-5} Ω·cm^2 for $6H$- and $4H$-SiC, respectively). To discover the possible roles of the phases, the interface, and alternating properties of the underlying SiC layer, Pt/Si contacts were removed and replaced by nickel contacts. Similar to the case of nickel silicide contacts to n-type SiC described above in the Section 4.1, the newly deposited Ni contacts were ohmic as deposited at room temperature and showed very little change in the characteristics relative to those of the original PtSi contacts. It was suggested that the changes in the underlying SiC layer created during the annealing process are responsible for the ohmic behaviour rather than the creation of PtSi phase itself [178]. In conclusion, the slight improvement of Pt based contacts in terms of contact resistivity and morphology in comparison to Ti/Al based contacts was obtained at the cost of more complicated processing which included silicon deposition at high temperature and pattering deposited films by wet chemical etching.

The second reason for searching alternative contact metallization schemes is that the Al-based ohmic contacts were considered as unreliable for commercial high temperature and high-power device applications due to aluminum's low melting temperature and high driving force for oxidation [136]. There are various aluminum free contacts and one of them, based on PtSi, is described above. Another example which is worth to mention is a cobalt silicide contact which is referred for many years as the contact with the lowest resistivity (3.6×10^{-6} Ω·cm^2) [165]. These contacts were obtained by a subsequent e-beam evaporation of 50 nm thick cobalt and silicon (160 nm) on $6H$-SiC epitaxial layers (1 μm; N_A=2×10^{19} cm^{-3}) followed by a two-step PDA: the first step at 500 °C for 5 hours to form

CoSi and the second step at 900 °C for 2 hours to form an ohmic contact. Fabricated contacts revealed very unusual behaviour, namely their resistance increased with both temperature and current density. It should be noted that the 120 nm thick pure Co film processed by the same way demonstrated contact resistivity of only 1×10^{-3} $\Omega\cdot cm^2$. These Co/Si contacts have not received any further development, most probably due to very high thermal budged required for their formation.

The third objective for searching contact metallization schemes alternative to the Al/Ti based contacts is to find a metal or a combination of metals which can form ohmic contacts simultaneously to both n- and p-type SiC. These contacts may greatly reduce a process complexity of some devices where both contacts to n- and p-type SiC have to be formed on the same wafer side (e.g. thyristors, planar JFETs, TI-VJFETs). One of the possible schemes is an epitaxially grown titanium carbide which was described above in Section 4.3. It provided contact resistivities of about 5×10^{-5} $\Omega\cdot cm^2$ on both n-type (N_D=$1.3\times10^{19}cm^{-3}$) and p-type (N_A>$1\times10^{20}cm^{-3}$) $4H$-SiC epi-layers at the cost of more complicated processing.

Usually, low resistivity of contacts to n-type SiC is more important in switching bipolar devices than the resistivity of contacts to p-type SiC. For instance, JFETs efficiency critically depends on source and drain ohmic contacts resistivities because of high channel current density while the resistivity of ohmic contacts to p-type SiC is not so critical due to low gate current density and is important only in transient processes. For this reason, it makes sense to check the applicability of a standard or slightly modified nickel silicide metallizations as an ohmic contact to p-type SiC. It was done by Ito et al. [71]. They investigated Al/Ni bilayers with different Al contents as an ohmic contact to both n- and p-type SiC with N_N= 1.3×10^{19} cm^{-3} and N_{Al}=7.2×10^{18} cm^{-3}, respectively. The aluminum films with thicknesses ranged from 0 to 10 nm were evaporated on $4H$-SiC epitaxial layers and covered by 50 nm thick nickel. The contacts were annealed at 1000 °C for 5 min in UHV. Ohmic contacts to n-type SiC were found to be formed when the Al-layer thickness was less than about 6 nm while ohmic contacts to p-type SiC were observed for an Al-layer thickness greater than about 5 nm. The contacts with an Al-layer thickness in the range of 5 to 6 nm exhibited ohmic behavior to both n- and p-type SiC, with a specific contact resistance of 1.8×10^{-4} $\Omega\cdot cm^2$ and 1.2×10^{-2} $\Omega\cdot cm^2$ for n- and p-type SiC, respectively. Very high contact resistivity to p-type SiC in this study most probably was caused by the relatively low doping level of $4H$-SiC layer (N_{Al}=7.2×10^{18} cm^{-3}). Indeed, Vivona et al. [61] reported ohmic contacts with ten times lower contact resistivity (ρ_C=1×10^{-3} $\Omega\cdot cm^2$) formed by PDA at 1000 °C for 1 min of pure nickel layer evaporated on $4H$-SiC with N_A=1×10^{20} cm^{-3} and Wang et al. [149] reported ohmic contacts with

ρ_C=2×10^{-5} $\Omega \cdot$cm^2 formed by PDA at 850 °C for 1 min of pure nickel layer evaporated on 4H-SiC with N_A=1×10^{21} cm^{-3}.

5.4   Ohmic contacts to heavily doped p-type SiC

All ohmic contacts to p-type SiC demonstrated strong decreasing of contact resistivity with temperature, except of CoSi contacts as discussed above. This dependence is always well fitted by the TFE model. Vivona *et al.* [148] extracted barrier heights in different ohmic contacts formed on Al implanted 4H-SiC layers with the same doping level (x_S=0.3 μm, N_{Al}=1×10^{20} cm^{-3}). The barrier heights ranged from 0.46 eV for Ti$_3$SiC$_2$ contacts to 0.75 eV for Ni$_2$Si contacts. These ϕ_B values are slightly lower than the one estimated in Section 2 and shown by curve 3 in Fig. 2 and this discrepancy may attributed to the interface roughness, additional surface states and alternation of SiC properties under contact as a result of PDA in real contacts. This roughly exponential temperature dependence of contact resistivity has to disappear at doping levels above 1×10^{21} cm^{-3} where the dominant mechanism of a charge carrier flow through Schottky barrier contacts to p-type 4H-SiC switches from TFE to FE (see Fig. 4) and further decrease of contact resistivity may be expected. That is why the formation of heavily doped SiC regions under contacts is of high interest.

The maximum concentration of aluminium atoms that can be incorporated into the SiC crystal lattice and play the role of electrically active impurities depends on the methods of epitaxy or doping. The highest concentration of aluminium atoms ever incorporated into epitaxial silicon carbide, on the order of 2.0×10^{21} cm^{-3}, was obtained by doping during epitaxy of 6H-SiC, using the sublimation sandwich method [28]. The epitaxial layers grown in this way had the highest crystal quality and all aluminium atoms electrically activated. Very high growth temperature exceeding 2000 °C and impossibility of doping control during the growth process precludes an application of this method for formation of contact layers in real device structures.

The maximum concentration of aluminium that can be attained for SiC doping during the epitaxial growth by CVD is roughly 2×10^{20} cm^{-3} [179]. It was shown that practically all Al atoms were electrically activated in layers grown at temperature of 1550 °C. However, triangular defects interpreted as inclusions of the 3C-SiC polytype were observed in the layers with highest concentrations. These inclusions can cause non-uniform current density, local overheating and consequently can affect the reliability and lifetime of the contacts. That is why the doping of commercially available epitaxial wafers is limited by 2×10^{19} cm^{-3}.

Heavily doped layers of p-type 4H- and 6H-SiC can be grown by low-temperature liquid-phase epitaxy (LTLPE) [27]. The growth temperature is relatively low and lies in the range 1100–1200 °C. The concentration of aluminium atoms in layers grown by LTLPE was found to be 7×10^{20} cm^{-3}, which is lower than the highest doping level obtained by the sublimation sandwich method but exceeds the maximum doping level obtained for CVD. Ohmic contacts with $\rho_C \sim 1 \times 10^{-4}$ $\Omega \cdot$cm^2 were fabricated on these layers [180], [181] but contacts with the same resistivity and formed directly on commercial epitaxial layers were reported by many authors (see Table 6 and Table 7).

Heavily doped contact layers can also be obtained by ion implantation of aluminium followed by post-implantation annealing. Tone *et al.* [133] formed ohmic contacts to Al implanted 6H- and 4H-SiC layers with thickness of 0.22 μm and N_{Al} ranged from 8×10^{19} cm^{-3} to 2×10^{21} cm^{-3}. The implantation was performed at room temperature and at 600 °C through 50 nm thick Si$_3$N$_4$ or SiO$_2$ capping layers to prevent an unwanted drop of the concentrations near the surface. Post implantation annealing was performed at 1500-1550 °C for 30 minutes in argon ambient. The samples were placed face-to-face in a graphite crucible with SiC to prevent surface deterioration and the PIA temperature was limited by step bunching observed at higher temperatures. The contacts were made of thermally evaporated 500 nm thick Al annealed at 950 °C for 5 min in Ar. The minimum resistivity of 3.3×10^{-5} $\Omega \cdot$cm^2 was measured in contacts to 6H-SiC with $N_{Al} = 8 \times 10^{20}$cm^{-3}. Contacts to 4H-SiC revealed the minimum resistivity of 6.1×10^{-5} $\Omega \cdot$cm^2 at $N_{Al} = 6 \times 10^{20}$ cm^{-3}. In both polytypes, the implanted regions were formed by co-implantation of carbon and aluminum at room temperature. Further increase of N_{Al} up to 2×10^{21} cm^{-3} did not result in decreasing of contact resistivity. These values of ρ_C are among the lowest published data but still significantly higher than the estimation of minimum contact resistivity made in Section 2. There are several possible reasons for this discrepancy. First, the PIA temperature used in this study was much lower than the temperature required for complete activation of implanted aluminium which is about up to 1800 °C [182]. Second, the implanted layers used in this study were not of the best quality. Indeed, they had a bulk resistivity of about 0.22 $\Omega \cdot$cm at $N_{Al} > 10^{21}$cm^{-3} which is of the same level as was measured in commercial CVD grow layers with $N_A = 2 \times 10^{19}$cm^{-3}. For comparison, the CVD grown p-type 6H-SiC layers with $N_A = 5 \times 10^{20}$ cm^{-3} exhibited the bulk resistivity of 0.042 $\Omega \cdot$cm [183] and $\rho_B = 0.02$ $\Omega \cdot$cm was measured in the LTLPE grown 4H-SiC layer with $N_{Al} = 2 \times 10^{20}$ cm^{-3} [180]. Finally, the third obvious reason of relatively high contact resistivities is the use of pure aluminum as a contact metallization which is not the best choice as it was discussed above in Section 5.1. Nevertheless, this study of ohmic contact to p-type SiC layers with doping levels close to the solubility limit published by Tone *et al.* [133] remains to be the only detailed publication on this subject, to the best of our knowledge.

6.   Compatibility of ohmic contacts formation with SiC device processing

The main drawback of ohmic contacts to silicon carbide is a high thermal budged required for their formation. This significantly hampers fabrication of SiC devices, especially devices with Schottky barrier contacts or MOS structures which may be damaged by following annealing of ohmic contacts. On the other hand, ohmic contacts formed before oxide growth or deposition can be damaged by high temperature processing in an oxidizing atmosphere. Furthermore, SiC oxidation or sintering a deposited oxide in presence of metal containing contacts can significantly deteriorate oxide properties. Therefore, SiC device processing flow has to be carefully designed to overcome these problems.

As it was discussed in previous sections, there are several methods to form ohmic contacts without high temperature post deposition annealing. For example, as-deposited metals can form low-resistivity ohmic contacts to heavily doped n-type SiC formed by phosphorus implantation [130] but the implantation itself requires post implantation annealing with thermal budged even higher than PDA of ohmic contacts. Therefore, it has to be done before other temperature sensitive processing steps. Another way to avoid the PDA step is to grow titanium carbide which forms ohmic contacts with decent resistivity to both n- and p-type SiC [58] but this process has to be carried out at 500 °C and this temperature is high enough to alter properties of Schottky contacts and MOS structures.

Ohmic contacts may be formed at first and then protected by silicon nitride to do not be damaged by following high temperature processing in an oxidizing atmosphere and do not contaminate grown or deposited oxides [184]. This approach was successfully used to fabricate 4H-SiC MOSFETs with increased mobility [185]. Nickel silicide ohmic contacts to n-type SiC were protected by 100 nm thick Si_3N_4 layer grown by PECVD before formation of a gate stack consisting of a thin SiO_2 layer was grown at 600 °C for 3 min by RTP in a dry oxygen ambient and 40 nm thick Al_2O_3 grown by ALD at 200 °C. This contact protection method was not tested for compatibility with conventional SiC thermal oxidation process and certainly is not compatible with high-temperature oxidation since the melting point of Ni_2Si is about 1250 °C.

Presumably, so called secondary contacts (discussed in Sections 4.1 and 5.3) can be used to combine high temperature annealing required for ohmic contacts formation with other SiC processing steps sensitive to high temperature and metal contamination, like gate oxide formation in MOSFETs. These contacts are formed by (1) the first deposition of a single metal (e.g. Ni) or multilayer (e.g. Si/Pt [178], Ti/Al [132]) film; (2) high temperature annealing resulted in alternation of underlying SiC layer; (3) removal of the solid state chemical reaction products to expose underlying SiC substrate; (4) second deposition of a metal layer at low temperature. These secondary contacts were demonstrated to be ohmic as-deposited on n-type [57], [123], [124] and p-type SiC [132], [178] and remained stable when subjected to thermal stress at 300 °C in air for 1400 hours [125]. The device fabrication process including temperature sensitive steps performed before the second metal deposition was patented [186] but there are no publications on fabrication of SiC devices utilizing this approach and nor their characterization.

Development and commercialization of high-voltage, high-power SiC Schottky diodes in the last decade brought to the forefront the problem of ohmic contacts formation by processing with reduced thermal budged. Indeed, SiC high-power devices with vertical current flow geometry have a drift layer of roughly 10 μm per 1 kV of blocking voltage (about 10 times thinner than one for silicon devices). For example, SiC Schottky diodes with blocking voltages about 1200 V have to have a drift layer of about 10 μm thickness with $N_D \sim 10^{16} cm^{-3}$. They are formed on substrates with a typical thickness of 350 μm and doping levels below $10^{19} cm^{-3}$. The electron mobility in SiC substrate is about $100 cm^2(V \cdot s)^{-1}$ which is ten times lower than the electron mobility in an epitaxial layer. It is easy to estimate that about 25% of total device resistance is the substrate resistance in this case. In case of 650 V diodes where the drift layer is about 5 μm the resulting substrate resistivity is about 70% of the total device resistivity [187]. Furthermore, the substrate contribution in total resistance is much higher in power devices requiring to be fabricated on p-type substrates like thyristors which are currently under development. Thinner SiC substrates cannot be used in device processing due to a high risk of wafer breakage and deformation. As a consequence, a SiC substrate has to be thinned either by grinding or by etching at the last processing steps followed only by the formation of backside ohmic contacts and their enforcement by a solderable metal stack. Conventional post deposition annealing as well as ion implantation with post implantation annealing cannot be used in this case because they destroy the device structure formed on the front side of the wafer. The solution was found in laser annealing of back-side contacts.

6.1   Laser annealing of back-side ohmic contacts

To minimize the thermal budget applied to the already finished device structures at the substrate front side, backside contacts can be formed by a contact layer deposition followed by pulse laser annealing (PLA). Pulse laser annealing is a thermally non-equilibrium process. A contact layer is irradiated with a laser beam and its temperature is raised due to photon absorption. During the pulse time (t_P), the heat penetrates for the depth of x_T from the surface:

$$x_T = 2\sqrt{\frac{t_p \lambda}{c\kappa}} \tag{39}$$

where λ is the SiC heat conductance (4.9 W/cm/K [13]); c is SiC specific heat capacity (0.69 J/g/K [13]) and κ is the SiC density (3.21 g/cm^3). By substituting these parameters in Eq. 39, it can be rewritten as:

$$x_T\,(\mu m) = 0.94\sqrt{t_p(ns)} \tag{39a}$$

The overheating temperature of the contact and SiC layer for the depth of about x_T during a single pulse can be roughly estimated as:

$$\Delta T = \frac{E_P}{x_T c\kappa} \tag{40}$$

By substituting x_T from Eq. 39 and the SiC parameters in Eq. 40, it can be rewritten as:

$$\Delta T(K) = 4800\frac{E_P(J/cm^2)}{\sqrt{t_p(ns)}} \tag{40a}$$

The overheating temperature of silicon carbide substrate at the distance of more than x_T from the contact is defined by heat spreading and by the average power of laser irradiation which can be kept low to avoid damaging the device structures on the substrate front side.

Ohmic contacts were formed by pulse laser annealing to both n- and p-type SiC. Selected results published over the last two decades are summarized in Table 8. It has to be noticed that all specific contact resistances given in Table 8 were measured by TLM on thick substrates without TLM structure termination and have to be considered only as a rough estimations. Nevertheless, it is clear that the contacts formed by PLA demonstrated resistivities comparable with values reported for the contacts fabricated on SiC layers with similar doping levels by conventional post deposition annealing.

A SiC substrate can be thinned down to the thickness of about 110 µm to remain strong enough for further processing and free of hairline cracks [187] while the heat penetration depth for t_p=200 ns (the maximum pulse length listed in Table 8) is only about 13 µm.

Eq. 40 gives reasonable ΔT values of about 1500 K (e.g. ΔT= 1360 K for t_p=200 ns and E_P=4 J/cm^2) for all contacts listed in Table 8 except of the one with PLA by Nd-YAG laser. This discrepancy may be explained by the fact that the estimation by Eq. 40 assumes that all pulse energy is absorbed by the contact while the proportion of absorbed energy strongly depends on metallisation material, surface roughness and laser wavelength. For example, the reflection factor for pure highly polished aluminium is 80-87% and only 50-60% for highly polished nickel. Furthermore, the contacts with PLA by Nd-YAG laser are the only contacts through that one listed in Table 8 which were annealed using the infrared irradiation and were formed on the as-grown surface of epitaxial layers while other contacts were deposited on the back side of substrates which has much higher roughness.

Table 8. Laser annealed ohmic contacts to SiC.

Ref.	ρ_C, $(10^{-4}\,\Omega\cdot c m^2)$	Metal (nm)	Substrate, doping (cm^{-3})	Laser, wavelength (nm)	Pulse number or scanning speed	Pulse length and energy density $(ns\times J/cm^2)$	Comments
[188]	0.43	Ti/W (10 or 30/10)	n 6H-SiC bulk; 1.5×10^{18}	KrF (248)	200	20× (1.5 or 2)	in vac. 2×10^{-6} torr
[188]	10	Ti/Al (10/60)	p 6H-SiC bulk; 1.5×10^{18}	KrF (248)	100	20×1.5	in vac. 2×10^{-6} torr
[189]	0.45-0.77	Ti, Ni, Pt, Au (200)	n 6H-SiC epi 4×10^{18}; Si-face	Nd-YAG (1006)	6	8×30	in air; 5 μm n-epi on n-substr.; ρ_C min for Ti
[187, 190]	n/r	Si/Ni, Ti, Co, Mo (15-20)	n-type	XeCl (310)	1	200×(3-4)	2 deposited layers: doped Si and metal
[191] [192]	5.3	Ni(50) Nb(50)	n 4H-SiC bulk; 1×10^{18}	n/r (355)	455 mm/s	45×2.25	in N_2 flow; TLM 100×800 μm
[193]	4	Ti (75)	n 4H-SiC bulk; 1×10^{18}	n/r (355)	455 mm/s	45×2.25	in Ar flow; TLM 100×800 um

Ohmic contacts to silicon carbide fabricated with pulse laser annealing were reported for the first time by Ota *et al.* [189]. They irradiated 200 nm thick metal films deposited on *n*-type 6*H*-SiC epitaxial layers with carrier density of 4.2×10^{18} cm^3 using a pulsed Nd:YAG laser with a pulse duration of 8 ns, a fluence of 30 J/cm^2 and 6 pulses of total irradiation. The PLA process was carried out in vacuum and contacts formed by PLA of various metals were investigated. The contact resistivity was lower for Ti, Pt, Ni, Au in that order consistent with the surface roughness improvement. The lowest resistivity of 4.5×10^{-5} $\Omega\cdot$cm^2 was measured in titanium contacts.

Nakashima *et al.* [188] demonstrated ohmic contacts to 4*H*-SiC formed by PLA using a KrF laser with a pulse duration of 20 ns, a fluence up to 2 J/cm^2 and pulse number up to 200. The contacts were formed by PLA in vacuum of two layer metallisations: Ti (10 or 30 nm thick) covered by 10 nm thick W for *n*-type and Ti (10 nm) /Al (60 nm) for *p*-type 4*H*-SiC. Laser fluence was optimized to reduce the surface roughness of annealed contacts. A very small intermixing of the metals with the SiC substrate has been observed in contrary to that of the same contact schemes annealed in a conventional furnace. After PLA, the Ti/W contacts had the surface roughness of 10 nm which was almost the same as the one just after the deposition. Creation of both tungsten carbide and titanium carbide was detected by XPS. The minimum measured contact resistivity was $4.3 \times 10^{-5} \, \Omega \cdot cm^2$. The Ti/Al contacts to p-type $4H$-SiC had the same surface roughness of 10 nm, titanium carbide layer detected by XPS and contact resistivity of $1 \times 10^{-3} \, \Omega \cdot cm^2$.

Infineon has patented the pulse laser annealing of back side contacts to n-type SiC [190], [187]. They proposed two procedures corresponding to two distinct mechanisms of contact formation. In the first mechanism, the laser irradiation was used to vaporize silicon from the SiC surface and leave thus, on the SiC surface, a conductive sp^2 carbon layer, the latter being reinforced by a metal multilayer like Ti/Ni/Ti/Ag. The carbon layer formed an ohmic contact to SiC. It was highlighted that the crucial point was to optimize the PLA process to form just the conducting carbon layer but not clusters of carbon material having a detrimental result in the ohmic character of the contact. In the second mechanism, the laser action helped in the creation of metal silicide by reaction between two deposited layers: a doped polycrystalline or amorphous Si layer and a metal (either Ni, Ti, Co, Mo) layer. The total thickness of silicon and metal layer was below 50 nm with metal thickness ranged from 15 to 20 nm. The contacts were annealed by XeCl laser (310 nm) with pulse length of 150 ns and variable pulse energy. The carbon released during the interfacial reaction between SiC and excess metal (Ni) in the deposited layer was found to be very fine dispersed in the resulting NiSi matrix, in contrast to relatively large carbon clusters formed by conventional PDA. It was supposed that this finer grained structure should lead to better integrity of the contact metal stack. This PLA process was used to form back-side contacts in SiC Schottky diodes after substrate thinning to 110 μm. The same method was proposed for use in front side contacts, particularly for the formation of merged p-n and Schottky diodes by using reticles to irradiate specific areas.

7.   Protection and overlaying ohmic contacts to SiC

The maximum operation temperature of silicon carbide devices is limited by the Debye temperature of SiC or by the temperature at which the intrinsic carrier concentration in SiC is the same as in silicon at 200 °C (5.3×10^{13} cm^{-3}), depending which one is lower. Debye temperature in $4H$-SiC is 1027 °C [13] and the intrinsic carrier concentration reaches 5.3×10^{13} cm^{-3} at 865 °C and this value should be considered as the maximum operating temperature of SiC devices. In practice, it is limited by thermal stability of dielectrics and metallizations used in a SiC device. It is worth noting that commercial

SiC MOSFETs and Schottky diodes are designed for operating junction and storage temperature only up to 175 °C [194] and significant degradation of nickel based source-drain contacts in commercial MOSFETs was found after aging at 300 °C in air for about 100 hours [195].

Standard ohmic contacts to SiC which are Ni and Al/Ti for *n*- and *p*-type SiC, respectively, are formed at temperatures exceeding 950 °C and expected to be thermally stable at lower temperatures. Indeed, Ni_2Si based contacts (formed by PDA of e-beam evaporated Ni at 950 °C for 10 minutes) remained stable after ageing for more than 100 hours in nitrogen at 500 °C [68]. The Ti_3SiC_2 based contacts to *p*-type SiC (formed by PDA of sputtered TiAl alloys) were stable and reliable after ageing tests performed at 600 °C under Ar atmosphere up to 400 hours [43]. However, ohmic contacts in real devices have to be coated with additional capping layer for interconnection or wire bonding. Materials of this capping layer (usually gold or platinum for high temperature applications) may diffuse into contact body to the interface with silicon carbide and change contact properties. Furthermore, long-term stability of ohmic contacts in vacuum or inert ambient does not mean the same in air atmosphere due to possible oxidation of contact and silicon carbide. Metals and oxygen can diffuse easily through Ni_2Si and Ti_3SiC_2. These contacts have to be protected with diffusion barrier or an alternative contact metallisation which can serve simultaneously as an ohmic contact and as a diffusion barrier (for example titanium carbide) should be considered. Consequently, a diffusion barrier that can prevent diffusion of oxygen and metals of capping layer into the contacts is essential for the development of SiC devices for high-temperature operation.

The major requirements for a suitable barrier layer are: (1) the barrier layer should not react with the underlying contact or capping metal; (2) it should have low electrical resistivity; (3) it should show good adhesion to both contacts and capping metals; (4) it is desirable to be amorphous to reduce a number of defects such as voids or grain boundaries. Nitrides of refractory metals such as TiN, TaN, and WN, carbides such as TiC and TaC, silicides such as $TaSi_2$ as well as metal alloys were tested as diffusion barriers for protection of ohmic contacts to silicon carbide due to their high stability and good electrical conductivity. Some most representative examples of barriers of each kind are discussed below in this section and more detailed review can be found in [9].

Baeri *et al.* [196] studied WTi(10%) sputtered alloy as a diffusion barrier. $Au/TiW/Ni_2Si$ (100/600/200 nm) contacts to *n*-type 6*H*-SiC demonstrated stable ohmic characteristics during ageing up to 450 °C for 100 hours in pure oxygen atmosphere. Strong interdiffusion and oxygen diffusion were observed at temperatures exceeding 500 °C and the contacts degraded just after 3 hours at 600 °C. On the other hand, it was shown [69] that a WTi alloy sputtered directly on a *n*-type 4*H*-SiC forms ohmic contact with

ρ_C=1.5×10^{-5} Ω·cm^2 after PDA at 950 °C for 30 minutes. After protection with e-beam evaporated 30 nm thick Ti adhesion layer and 300 nm thick platinum, it remained stable in air at 600 °C for 250 hours.

Another alloy tested as a diffusion barrier was reported by Wang *et al.* [149]. A 200 nm thick diffusion barrier was formed by co-sputtering of tantalum and rubidium on top of Ni and Al/Ni contacts to *n*- and *p*-4*H*-SiC. It was covered by 700 nm tick Au capping layer. Both contacts to *n*- and *p*-4*H*-SiC demonstrated resistivity of about 10^{-5} Ω·cm^2 which remained unchanged after aging at 350 °C for 3000 hours in air.

Carbides of titanium and tantalum are very good diffusion barriers and can form ohmic contacts to *n*-type (TiC, TaC) and *p*-type SiC (TiC) combining contact and diffusion barrier in a single layer. Jang *et al.* [108] investigated tantalum carbide (TaC) as an ohmic contact to *n*-type SiC. A 100 nm thick TaC film was sputtered at 200 °C on *n*-type epi-layer (0.5 μm; 8×10^{18}cm^{-3}) grown in Si-face of *n*-type 6*H*-SiC substrate. The PDA at 1000 °C for 15 min in vacuum was required to get ohmic contacts with ρ_C=3×10^{-5} Ω·cm^2. These contacts demonstrated very good thermal stability when tested with W/WC (100 nm/50 nm) protection layer. They showed no degradation after ageing in vacuum at 600 °C for 1000 hours and at 1000 °C for 500 hours (note that this temperature exceeds the maximum operating temperature of SiC devices).

Titanium nitride diffusion barriers on Ti and Ni based ohmic contacts to *n*-type 4*H*-SiC (3×10^{19} cm^{-3}) were studied by Daves *et al.* [70]. The contacts were formed by PDA in Ar at 1050 °C and 950 °C of 30 nm thick Ti and 50 nm thick Ni, respectively. Then, the multilayer Ti(adhesion)/TiN/Pt/Ti(adhesion) (20/10/150/20 nm) barrier film was sputtered and coated by PECVD grown a-SiO$_x$/a-SiC stack (250/250 nm). Ti ohmic contacts with initial resistivity of 6×10^{-5} Ω·cm^2 showed the best performance. They remained ohmic after 500 hours of aging at 600 °C in dry and wet air (10% moisture) ambient without significant degradation of morphological properties and with gradual resistivity increase to 2.3×10^{-4} Ω·cm^2. This contacts withstood temperature cycling from 100 to 700 °C in wet air for 100 hours. On the other hand, Ni-based contacts with initial ρ_C=3×10^{-5} Ω·cm^2 degraded very quickly during the first burn-in at 600 °C. Provided TEM images clearly shown that the TiN barrier layer was not thick enough to completely prevent platinum diffusion into the contact. Indeed, the titanium carbide layer formed during PDA of Ti based contacts served as an additional barrier to stop platinum diffusion and to remain the interface between the contact and silicon carbide unaltered. This resulted in a good thermal stability demonstrated by these contacts. In contrast, Ni$_2$Si contact did not provide an additional diffusion barrier and allowed significant platinum diffusion through the TiN barrier layer and Ni$_2$Si contact layer down to the contact interface with SiC. Platinum reaction with SiC was detected in Ni based contacts explaining the fast degradation of these contacts. It should be noted that the contacts were protected by additional a-SiO$_x$/a-SiC coating against the in-diffusion of oxygen into the contact metallization.

The most systematically studied metallization scheme of capping layer is Pt with TaSi$_x$ diffusion barrier. Virshup *et al.* [197] investigated the thermal stability of 4*H*-SiC/Ni/TaSi$_x$/Pt ohmic contacts. A 100 nm thick thermally evaporated Ni was annealed at 950 °C for 5 minutes in argon and then covered by sputtered 50 nm thick TaSi$_x$ and 150 nm thick Pt. The contacts remained ohmic with gradually increasing resistivity from 3.4×10^{-4} to 2.8×10^{-3} $\Omega \cdot cm^2$ during the aging test for over 1000 hours in air at 300 °C. The contacts heated at 500 °C and 600 °C, however, showed larger increases in specific contact resistance followed by nonohmic behavior after 240 and 36 hours, respectively. It was proven by AES profiling that the loss of ohmic behavior occurred when the entire tantalum silicide layer had oxidized.

The highly durable *n*-ohmic metallization scheme based on tantalum silicide was developed by Okojie *et al.* [78], [78, 198]. The Ti(100 nm)/TaSi$_2$(400 nm)/Pt(300 nm) multilayer stack was deposited by magnetron sputtering on *n*-type 6*H*- and 4*H*-SiC epitaxial layers with $N_D=7 \times 10^{18}$ cm^{-3} and $N_D=2 \times 10^{19}$ cm^{-3}, respectively. After annealing at 600 °C for 30 min in nitrogen, platinum silicide had formed on the topmost surface via the solid state chemical reaction:

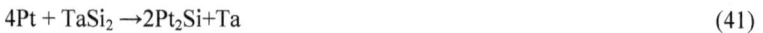

$$4Pt + TaSi_2 \rightarrow 2Pt_2Si + Ta \qquad (41)$$

At the Ti/SiC reaction zone, XTEM analysis indicated the formation of titanium silicide as the reaction product in intimate contact with SiC. The resistivity of 1.7×10^{-4} $\Omega \cdot cm^2$ and 4.5×10^{-4} $\Omega \cdot cm^2$ was measured in contacts to 6*H*- and 4*H*-SiC, respectively.

After contacts annealing for the first 100 hours at 600 °C in air, the platinum silicide on the surface was found reacted with oxygen to form an about 40 nm thick protective layer of silicon oxide. It was supposed that this is the layer responsible for the protection of the contact from further oxidation. The contact resistivity reduced down to about 4.5×10^{-5} $\Omega \cdot cm^2$ and remained stable up to 1000 hours of further ageing. Contacts with different layer thicknesses were also tested. It was found that the 200 nm TaSi$_2$ layer is not thick enough to prevent titanium out-diffusion and oxidation. The main reason of contact degradation was found to be the oxidation through the edges of Ti containing layers. It was concluded that the edges have to be protected as it was done later by Neudeck *et al.* [199, 200] with Si$_3$N$_4$ deposited by reactive sputtering a high-purity silicon target at 7 mTorr chamber pressure, N$_2$ injected near the middle of the chamber and Ar injected at a sputter gun.

Advancing Silicon Carbide Electronics Technology I   Materials Research Forum LLC
Materials Research Foundations **37** (2018)   doi: http://dx.doi.org/10.21741/9781945291852

This metal stack was used by Neudeck *et al.* [199, 200] to fabricate 6*H*-SiC lateral JFETs which demonstrated a stable operation up to 6000 hours at 500 °C. The same metallization scheme was used for contacts to *n*-type source/drain implanted regions as well as to the heavily *p*-type doped ($>10^{20}$ cm^{-3}) gate epilayer. A second layer of the same TaSi$_2$/Pt composition was used as an interconnection. Surface was passivated by 1.1 μm thick Si$_3$N$_4$ layer deposited by reactive sputtering and overlapping the contact edges.

The same contact scheme was used to fabricate 6*H*-SiC MESFETs [201]. Both Schottky and ohmic contacts were formed by annealing of Ti/TaSi$_2$/Pt multilayer in nitrogen at 600 °C for 30 minutes. MESFETs were aged in air at 500 °C up to 2400 hours. Transistors were leaky initially and leakage increased after ageing but R_{DS} remained unchanged and the conclusion was made about the stability of ohmic contacts.

Very impressive results were published recently on using TiSi$_2$ for contacts protection and interconnection [202]. Digital and analog integrated circuits (ICs) based on 4*H*-SiC JFETs were fabricated using 50 nm sputtered Hf as ohmic contacts both to *p*-type 4H-SiC ($N_A>10^{20}$ cm^{-3}) and to phosphorus implanted *n*-type 4*H*-SiC (x_S=0.4 μm; $N_D>10^{20}$ cm^{-3}). A 200 nm thick magnetron sputtered TaSi$_2$ was used as a capping layer. Two levels of interconnect were made of 0.8 μm thick TaSi$_2$ with 1 μm thick SiO$_2$ as an interlayer dielectric. Finally, the circuits were protected by two 1 μm thick SiO$_2$ layers with 67 nm thick Si$_3$N$_4$ layer between them. Although no data about contact resistivity was published, the electrical characteristics of 11-stage ring oscillator ICs and operational amplifiers were presented. Following initial burn-in, important circuit parameters remain stable within 15% for more than 2000 hours of operation in air at 500 °C.

8.   Conclusion

It is more than forty years ago since the first SiC devices with ohmic contacts fabricated by conventional techniques were reported [172]. W. von Muench, and I. Pfaffeneder used post deposition annealing at 1100 °C of evaporated nickel and aluminium to form ohmic contacts to *n*-type CVD layers (N_D=2×10^{19} cm^{-3}) and *p*-type 6*H*-SiC Lely substrates (N_A=2×10^{18} cm^{-3}), respectively. This technique is still widely usable but significant progress has been made since that time.

Ohmic contacts to *n*-type SiC are developed to that point where the lowest published contact resistivity ($\sim3\times10^{-7}$ Ω·cm^2) is very close to its estimated lower bound ($\sim2\times10^{-7}$ Ω·cm^2). These contacts can be fabricated without high temperature post deposition annealing owing to the availability of high quality heavily doped silicon carbide formed by ion implantation. Several methods were developed to form ohmic contact with decent resistivity ($\sim1\times10^{-5}$ Ω·cm^2) to *n*-type SiC substrates

($N_D\sim2\times10^{18}\,\text{cm}^{-3}$) and epitaxial CVD layers ($N_D\sim2\times10^{19}\,\text{cm}^{-3}$). These methods are based on interfacial reactions between silicon carbide and deposited metals to form metal silicides, carbides and ternary compounds. Rapid thermal processing is widely used to control these reactions with high precision. Furthermore, development of high power pulsed lasers has made it possible to perform local contact annealing with low thermal budget annealing which significantly improved compatibility of ohmic contacts formation with device processing.

Ohmic contacts to p-type SiC are not as advanced as contacts to n-type SiC. First of all, the theoretically estimated minimum resistivity of contacts to p-type SiC ($\sim1\times10^{-6}\,\Omega\cdot\text{cm}^2$) is higher than the n-type SiC, simply because of the SiC material properties. The best published resistivity of contacts to p-type SiC ($\sim1\times10^{-5}\,\Omega\cdot\text{cm}^2$) is still far higher than this value. Furthermore, all ohmic contacts to p-type SiC require high temperature post deposition annealing or metal deposition at high temperature ($\sim600\,°C$) even when formed to SiC with doping level close to the impurity solubility limit. Nevertheless, decent ohmic contacts with $\rho_C < 1\times10^{-4}\,\Omega\cdot\text{cm}^2$ are routinely obtained on p-type SiC with $N_A > 1\times10^{19}\,\text{cm}^{-3}$. Epitaxial structures with this doping level are commercially available and there is no need in additional ion implantation to form these contacts.

It is worth to note that the main application area of ohmic contacts is currently silicon carbide power switching and rectifying devices. They have vertical geometry and are fabricated on 4H-SiC commercial substrates. These substrates have the minimum bulk resistivity of about $0.02\,\Omega\cdot\text{cm}$ which results in specific series resistance of about $2\times10^{-4}\,\Omega\cdot\text{cm}^2$ for substrates thinned down to 100 µm. The area of backside contacts in SiC power devices is roughly the same as the chip area while the area of the top contacts on the front side can be ranged from the same as the chip area in rectifying devices (e.g. p-i-n diodes) to about 1/10 of the chip area in switching devices (e.g. MOSFETs). Obviously, ohmic contacts to n-type SiC have sufficiently low resistivity ($\rho_C < 1\times10^{-5}\,\Omega\cdot\text{cm}^2$ are routinely obtained) to neglect their resistance in power SiC devices while further technology development is required for reproducible formation of ohmic contacts to p-type SiC with $\rho_C < 1\times10^{-5}\,\Omega\cdot\text{cm}^2$.

Remarkable results were obtained in the development of protection coatings and diffusion barriers for ohmic contacts to silicon carbide. Integrated circuits based on planar SiC JFETs (ring oscillators and operational amplifiers) capable to operate in oxidizing atmosphere at 500 °C for more than 2000 hours were demonstrated. In these devices, tantalum silicide was used as a material for diffusion barriers and interconnections and multilayer $SiO_2/Si_3N_4/SiO_2$ coating was used as a protection layer.

Beyond doubts, fabrication of ohmic contacts to silicon carbide is a rather matured technology but there are still many challenges which have to be addressed. Most of them are technological problems such as further improvement of contacts adhesion, reduction of PDA thermal budget and search for more efficient diffusion barriers. Ohmic contacts to both *n*- and *p*-type SiC with doping levels close to the impurity solubility limit remain to be not sufficiently studied up to now. Two other problems are more scientific than technological. The first one is to elucidate the changes in a thin SiC layer under contacts which resulted from the interfacial reaction with deposited metals during post deposition annealing. Many authors came to the conclusion that this alternating SiC property is the main factor leading to the change of contact behaviour from Schottky barrier to ohmic but there is no published data on material characterisation of this thin SiC layer under contact. The second challenge is to develop correct methods of contact resistivity measurements applicable for characterisation of ohmic contacts to silicon carbide. Indeed, the contact resistivity methods like CBKR and TLM are based on the transition line model and were developed for measurements on thin silicon layers. The higher a layer bulk resistivity is thinner layers are required to remain this model correctly applicable. The bulk resistivity of silicon carbide is about 20 times higher than the one of silicon and direct transfer of contact resistivity measurement methods from silicon to silicon carbide requires corresponding reduction of SiC layer thickness. Measurements by TLM of contact resistivities lower than 10^{-5} and 10^{-6} $\Omega \cdot cm^2$ in contact to *p*- and *n*-type SiC, respectively, require test structures to be formed on unrealistically thin SiC layers.

In conclusion, it is worth to recall that thirty years ago, Rhoderick and Williams noticed that the fabrication of ohmic contacts is more of an art than a science [10] and this statement still remains to be pertinent with respect to ohmic contacts to silicon carbide.

Acknowledgements

This work was supported by the UK EPSRC (Grant EP/L007010/1).

References

[1]   R. N. Hall, "Electrical Contacts to Silicon Carbide," *Journal of Applied Physics,* vol. 29, no. 6, pp. 914-917, 1958.

[2]   J. A. Lely, "Darstellung von Einkristallen von Silicium Carbid und Beherrschung von Art und Menge der eingebauten Verunreinigungen," *Ber. Dt. Keram. Ges.,* vol. 8, pp. 229, 1955.

[3]   Y. M. Tairov, and V. F. Tsvetkov, "Investigation of growth processes of ingots of silicon carbide single crystals," *Journal of Crystal Growth,* vol. 43, no. 2, pp. 209-212, 1978.

[4]   N. Kuroda, K. Shibahara, W. Yoo, S. Nishino, and H. Matsunami, "Step-Controlled VPE Growth of SiC Single Crystals at Low Temperatures," in Extended Abstracts of the 1987 Conference on Solid State Devices and Materials, 1987, pp. 227-230. https://doi.org/10.7567/SSDM.1987.C-4-2

[5]   G. L. Harris, G. Kelner, and M. Shur, "Ohmic contacts to SiC," *Properties of Silicon Carbide*, G. L. Harris, ed., p. 295, London, United Kingdom: INSPEC, the Institution of Electrical Engineers, 1993.

[6]   L. M. Porter, and R. F. Davis, "A critical review of ohmic and rectifying contacts for silicon carbide," *Materials Science and Engineering: B,* vol. 34, no. 2, pp. 83-105, 1995.

[7]   J. Crofton, L. M. Porter, and J. R. Williams, "The Physics of Ohmic Contacts to SiC," *physica status solidi (b),* vol. 202, no. 1, pp. 581-603, 1997.

[8]   F. Roccaforte, F. La Via, and V. Raineri, "OHMIC CONTACTS TO SIC," *International Journal of High Speed Electronics and Systems,* vol. 15, no. 04, pp. 781-820, 2005. https://doi.org/10.1142/S0129156405003429

[9]   Z. Wang, W. Liu, and C. Wang, "Recent Progress in Ohmic Contacts to Silicon Carbide for High-Temperature Applications," *Journal of Electronic Materials,* vol. 45, no. 1, pp. 267-284, 2016/01/01, 2016.

[10]   E. H. Rhoderick, and R. H. Williams, *Metal-Semiconductor Contacts*, 2 ed., Oxford: Clarendon Press, 1988.

[11]   D. K. Schroder, "Semiconductor Material and Device Characterization," John Wiley & Sons, Inc., 2005. https://doi.org/10.1002/0471749095

[12]   S. M. Sze, *Physics of Semiconductor Devices*, Second ed., p. 868, New York: Wiley, 1981.

[13]   Y. A. Goldberg, M. E. Levinshtein, and S. L. Rumyantsev, "Silicon Carbide," *Properties of Advanced Semiconductor Materials: GaN, AIN, InN, BN, SiC, SiGe*, M. E. Levinshtein, S. L. Rumyantsev and M. S. Shur, eds., New York: John Wiley & Sons, Inc. , 2001.

[14]   S. Y. Davydov, "On the electron affinity of silicon carbide polytypes," *Semiconductors,* vol. 41, no. 6, pp. 696-698, June 01, 2007. https://doi.org/10.1134/S1063782607060152

[15] M. Wiets, M. Weinelt, and T. Fauster, "Electronic structure of SiC(0001) surfaces studied by two-photon photoemission," *Physical Review B,* vol. 68, no. 12, pp. 125321, 2003.

[16] K. K. Ng, and R. Liu, "On the calculation of specific contact resistivity on <100> Si," *IEEE Transactions on Electron Devices,* vol. 37, no. 6, pp. 1535-1537, 1990. https://doi.org/10.1109/16.106252

[17] C. Persson, and U. Lindefelt, "Relativistic band structure calculation of cubic and hexagonal SiC polytypes," *Journal of Applied Physics,* vol. 82, pp. 5496-5508, 1997. https://doi.org/10.1063/1.365578

[18] W. M. Chen, N. T. Son, E. Janzén, D. M. Hofmann, and B. K. Meyer, "Effective Masses in SiC Determined by Cyclotron Resonance Experiments," *physica status solidi (a),* vol. 162, no. 1, pp. 79-93, 1997.

[19] N. T. Son, W. M. Chen, O. Kordina, A. O. Konstantinov, B. Monemar, E. Janzen, D. M. Hofman, D. Volm, M. Drechsler, and B. K. Meyer, "Electron effective masses in 4H SiC," *Applied Physics Letters,* vol. 66, no. 9, pp. 1074-1076, 1995. https://doi.org/10.1063/1.113576

[20] C. R. Crowell, "The Richardson constant for thermionic emission in Schottky barrier diodes," *Solid-State Electronics,* vol. 8, no. 4, pp. 395-399, 1965. https://doi.org/10.1016/0038-1101(65)90116-4

[21] G. Pensl, F. Ciobanu, T. Frank, M. Krieger, S. Reshanov, F. Schmid, and M. Weidner, "SiC MATERIAL PROPERTIES," *International Journal of High Speed Electronics and Systems,* vol. 15, no. 04, pp. 705-745, 2005. https://doi.org/10.1142/S0129156405003405

[22] A. Itoh, and H. Matsunami, "Analysis of Schottky Barrier Heights of Metal/SiC Contacts and Its Possible Application to High-Voltage Rectifying Devices," *physica status solidi (a),* vol. 162, no. 1, pp. 389-408, 1997.

[23] R. C. Jaeger, *Introduction to Microelectronic Fabrication,* New York: Addison-Wesley Publishing Company, 1993.

[24] Y. M. Tairov, and V. F. Tsvetkov, "Semiconductor Compounds AIVBIV," *Handbook on electrotechnical materials,* Y. V. Koritskii, V. V. Pasynkov and B. M. Tareev, eds., p. 728, Leningrad: Energomashizdat, 1988 [in Russian].

[25] A. Hallén, R. Nipoti, S. E. Saddow, S. Rao, and B. G. Svensson, "Advances in Selective Doping of SiC Via Ion Implantation," *Advances in Silicon Carbide*

Processing and Applications, S. E. Saddow and A. Agarwal, eds., pp. 109-153: Artech House, 2004.

[26]   S. W. Jones, *Diffusion in silicon*: IC Knowledge LLC, 2008.

[27]   S. V. Rendakova, V. Ivantsov, and V. A. Dmitriev, *Mater. Sci. Forum,* vol. 163, pp. 264-268, 1998.

[28]   Y. Vodakov, E. N. Mokhov, M. G. Ramm, and A. D. Roenkov, *Amorphous and Crystalline Silicon Carbide III [Springer Proceedings in Physics 56]* pp. 329, 1992. https://doi.org/10.1007/978-3-642-84402-7_50

[29]   R. S. Okojie, L. J. Evans, D. Lukco, and J. P. Morris, "A Novel Tungsten-Nickel Alloy Ohmic Contact to SiC at 900C," *Electron Device Letters, IEEE,* vol. 31, no. 8, pp. 791-793, 2010. https://doi.org/10.1109/LED.2010.2050761

[30]   M. Qin, V. M. C. Poon, and S. C. H. Ho, "Investigation of Polycrystalline Nickel Silicide Films as a Gate Material," *Journal of The Electrochemical Society,* vol. 148, no. 5, pp. G271, 2001. https://doi.org/10.1149/1.1362551

[31]   L. J. Chen, *Silicide Technology for Integrated Circuits*, Stevenage: Institution of Electrical Engineer, 2005.

[32]   Y.-J. Chang, and J. L. Erskine, "Diffusion layers and the Schottky-barrier height in nickel silicide—silicon interfaces," *Physical Review B,* vol. 28, no. 10, pp. 5766-5773, 1983.

[33]   H. Yu, X. Zhang, H. Shen, Y. Tang, Y. Bai, Y. Wu, K. Liu, and X. Liu, "Thermal stability of Ni/Ti/Al ohmic contacts to p-type 4H-SiC," *Journal of Applied Physics,* vol. 117, no. 2, pp. 025703, 2015.

[34]   A. Nino, T. Hirabara, S. Sugiyama, and H. Taimatsu, "Preparation and characterization of tantalum carbide (TaC) ceramics," *International Journal of Refractory Metals and Hard Materials,* vol. 52, pp. 203-208, 2015.

[35]   C. P. Kempter, and M. R. Nadler, "Thermal Expansion of Tantalum Monocarbide to 3020°C," *The Journal of Chemical Physics,* vol. 43, no. 5, pp. 1739-1742, 1965. https://doi.org/10.1063/1.1696999

[36]   J. B. William, "Work Function of Tantalum Carbide and the Effects of Adsorption and Sputtering of Cesium," *Journal of Applied Physics,* vol. 42, no. 7, pp. 2682-2688, 1971. https://doi.org/10.1063/1.1660608

[37]   U. Gottlieb, O. Laborde, O. Thomas, F. Weiss, A. Rouault, J. P. Senateur, and R. Madar, "Resistivity and magnetoresistance of monocrystalline $TaSi_2$ and VSi_2," *Surface and Coatings Technology*, vol. 45, no. 1, pp. 237-243, 1991.

[38]   I. Engström, and B. Lönnberg, "Thermal expansion studies of the group IV-VII transition-metal disilicides," *Journal of Applied Physics*, vol. 63, no. 9, pp. 4476-4484, 1988.

[39]   R. G. Wilson, and W. E. McKee, "Vacuum Thermionic Work Functions and Thermal Stability of TaB_2, ZrC, Mo_2C, $MoSi_2$, $TaSi_2$, and WSi_2," *Journal of Applied Physics*, vol. 38, no. 4, pp. 1716-1718, 1967. https://doi.org/10.1063/1.1709747

[40]   D. T. Morelli, "Thermal conductivity and thermoelectric power of titanium carbide single crystals," *Physical Review B*, vol. 44, no. 11, pp. 5453-5458, 1991.

[41]   S. Zaima, Y. Shibata, H. Adachi, C. Oshima, S. Otani, M. Aono, and Y. Ishizawa, "Atomic chemical composition and reactivity of the TiC(111) surface," *Surface Science*, vol. 157, no. 2-3, pp. 380-392, 1985.

[42]   M. Kadoshima, T. Matsuki, S. Miyazaki, K. Shiraishi, T. Chikyo, K. Yamada, T. Aoyama, Y. Nara, and Y. Ohji, "Effective-Work-Function Control by Varying the TiN Thickness in Poly-Si/TiN Gate Electrodes for Scaled High-k CMOSFETs," *IEEE Electron Device Letters*, vol. 30, no. 5, pp. 466-468, 2009. https://doi.org/10.1109/LED.2009.2016585

[43]   T. Abi-Tannous, M. Soueidan, G. Ferro, M. Lazar, C. Raynaud, B. Toury, M. F. Beaufort, J. F. Barbot, O. Dezellus, and D. Planson, "A Study on the Temperature of Ohmic Contact to p-Type SiC Based on Ti_3SiC_2 Phase," *IEEE Transactions on Electron Devices*, vol. 63, no. 6, pp. 2462-2468, 2016. https://doi.org/10.1109/TED.2016.2556725

[44]   M. Gao, S. Tsukimoto, S. H. Goss, S. P. Tumakha, T. Onishi, M. Murakami, and L. J. Brillson, "Role of Interface Layers and Localized States in TiAl-Based Ohmic Contacts to p-Type 4H-SiC," *Journal of Electronic Materials*, vol. 36, no. 4, pp. 277-284, 2007.

[45]   S. Y. Jiang, X. Y. Li, and Z. Z. Chen, "Role of W in W/Ni Bilayer Ohmic Contact to n-Type 4H-SiC From the Perspective of Device Applications," *IEEE Transactions on Electron Devices*, vol. 65, no. 2, pp. 641-647, 2018. https://doi.org/10.1109/TED.2017.2784098

[46] E. N. Denbnovetskaya, V. A. Lavrenko, I. A. Podchernyaeva, T. G. Protsenko, N. I. Siman, and V. S. Fomenko, "Electron work function and surface recombination of hydrogen for alloys of the system HfC-WC," *Soviet Powder Metallurgy and Metal Ceramics,* vol. 10, no. 4, pp. 289-291, April 01, 1971.

[47] A. L. Syrkin, J. M. Bluet, G. Bastide, T. Bretagnon, A. A. Lebedev, M. G. Rastegaeva, N. S. Savkina, and V. E. Chelnokov, "Surface barrier height in metal-SiC structures of 6H, 4H and 3C polytypes," *Materials Science and Engineering: B,* vol. 46, no. 1, pp. 236-239, 1997.

[48] A. Iton, O. Takemura, T. Kimoto, and H. Matsunami, "Barrier height analysis of metal/4H-SiC Schottky contacts," *Inst. Phys. Conf. Ser. No 142,* pp. 689, 1996.

[49] F. A. Padovani, and R. Stratton, "Field and thermionic-field emission in Schottky barriers," *Solid-State Electronics,* vol. 9, no. 7, pp. 695-707, 1966.

[50] C. R. Crowell, "Richardson constant and tunneling effective mass for thermionic and thermionic-field emission in Schottky barrier diodes," *Solid-State Electronics,* vol. 12, no. 1, pp. 55-59, 1969. https://doi.org/10.1016/0038-1101(69)90135-X

[51] A. Y. C. Yu, "Electron tunneling and contact resistance of metal-silicon contact barriers," *Solid-State Electronics,* vol. 13, no. 2, pp. 239-247, 1970.

[52] R. H. Cox, and H. Strack, "Ohmic contacts for GaAs devices," *Solid-State Electronics,* vol. 10, no. 12, pp. 1213-1218, 1967.

[53] G. K. Reeves, and H. B. Harrison, "Obtaining the specific contact resistance from transmission line model measurements," *IEEE Electron Device Letters,* vol. 3, no. 5, pp. 111-113, 1982. https://doi.org/10.1109/EDL.1982.25502

[54] H. Murrmann, and D. Widmann, "Current crowding on metal contacts to planar devices," *IEEE Transactions on Electron Devices,* vol. 16, no. 12, pp. 1022-1024, 1969.

[55] U. Haw-Jye, D. B. Janes, and K. J. Webb, "Error analysis leading to design criteria for transmission line model characterization of ohmic contacts," *IEEE Transactions on Electron Devices,* vol. 48, no. 4, pp. 758-766, 2001. https://doi.org/10.1109/16.915721

[56] A. N. Andreev, A. M. Strel'chuk, N. S. Savkina, F. M. Snegov, and V. E. Chelnokov, "An investigation of SiC-6H dinistor structures," *Pisma Zh. Tekh. Fiz.,* vol. 29, no. 6, pp. 1083-1092, 1995.

[57] I. P. Nikitina, K. V. Vassilevski, N. G. Wright, A. B. Horsfall, A. G. Oneill, and C. M. Johnson, "Formation and role of graphite and nickel silicide in nickel based

ohmic contacts to n-type silicon carbide," *Journal of Applied Physics,* vol. 97, no. 8, pp. 083709-083709-7, 2005. https://doi.org/10.1063/1.1872200

[58]   S. K. Lee, C. M. Zetterling, M. Östling, J. P. Palmquist, H. Högberg, and U. Jansson, "Low resistivity ohmic titanium carbide contacts to n- and p-type 4H-silicon carbide," *Solid-State Electronics,* vol. 44, no. 7, pp. 1179-1186, 2000.

[59]   S. Tanimoto, M. Inada, N. Kiritani, M. Hoshi, H. Okushi, and K. Arai, "Single Contact-Material MESFETs on 4H-SiC," *Materials Science Forum,* vol. 457-460, pp. 1221-1224, 2004. https://doi.org/10.4028/www.scientific.net/MSF.457-460.1221

[60]   H. Na, H. Kim, K. Adachi, N. Kiritani, S. Tanimoto, H. Okushi, and K. Arai, "High-quality schottky and ohmic contacts in planar 4H-SiC metal semiconductor field-effect transistors and device performance," *Journal of Electronic Materials,* vol. 33, no. 2, pp. 89-93, 2004.

[61]   M. Vivona, G. Greco, F. Giannazzo, R. Lo Nigro, S. Rascunà, M. Saggio, and F. Roccaforte, "Thermal stability of the current transport mechanisms in Ni-based Ohmic contacts on n- and p-implanted 4H-SiC," *Semiconductor Science and Technology,* vol. 29, no. 7, pp. 075018, 2014.

[62]   S. Tanimoto, N. Kiritani, M. Hoshi, and H. Okushi, "Ohmic Contact Structure and Fabrication Process Applicable to Practical SiC Devices," *Materials Science Forum,* vol. 389-393, pp. 879-884, 2002. https://doi.org/10.4028/www.scientific.net/MSF.389-393.879

[63]   S. Tanimoto, and H. Oohashi, "High-Temperature Reliable Ni_2Si-Based Contacts on SiC Connected to Si-Doped Al Interconnect via Ta/TaN Barrier," *Materials Science Forum,* vol. 615-617, pp. 561-564, 2009. https://doi.org/10.4028/www.scientific.net/MSF.615-617.561

[64]   W. Daves, A. Krauss, V. Häublein, A. Bauer, and L. Frey, "Enhancement of the Stability of Ti and Ni Ohmic Contacts to 4H-SiC with a Stable Protective Coating for Harsh Environment Applications," *Journal of Electronic Materials,* vol. 40, no. 9, pp. 1990-1997, 2011. https://doi.org/10.1007/s11664-011-1681-2

[65]   K. Vassilevski, S. K. Roy, N. Wood, A. B. Horsfall, and N. G. Wright, "Process Compatibility of Heavily Nitrogen Doped Layers Formed by Ion Implantation in Silicon Carbide Devices," *Materials Science Forum,* vol. 821-823, pp. 411-415, 2015. https://doi.org/10.4028/www.scientific.net/MSF.821-823.411

[66]   C. Arnodo, S. Tyc, F. Wyczisk, and C. Brylinski, "Nickel and molibdenum ohmic contacts on silicon carbide," *Inst. Phys. Conf.*, vol. 142, pp. 577, 1996.

[67]   R. Kakanakov, L. Kassamakova-Kolaklieva, N. Hristeva, G. Lepoeva, and K. Zekentes, "Thermally stable low resistivity ohmic contacts for high power and high temperature SiC device applications," *23rd International Conference on Microelectronics. Proceedings* vol. 1, pp. 205-208, 2002.

[68]   T. Marinova, A. Kakanakova-Georgieva, V. Krastev, R. Kakanakov, M. Neshev, L. Kassamakova, O. Noblanc, C. Arnodo, S. Cassette, C. Brylinski, B. Pecz, G. Radnoczi, and G. Vincze, "Nickel based ohmic contacts on SiC," *Materials Science and Engineering: B,* vol. 46, no. 1-3, pp. 223-226, 1997.

[69]   S.-K. Lee, E.-K. Suh, N.-K. Cho, H.-D. Park, L. Uneus, and A. L. Spetz, "Comparison study of ohmic contacts to 4H-silicon carbide in oxidizing ambient for harsh environment gas sensor applications," *Solid-State Electronics,* vol. 49, no. 8, pp. 1297-1301, 2005.

[70]   W. Daves, A. Krauss, V. Häublein, A. J. Bauer, and L. Frey, "Structural and Reliability Analysis of Ohmic Contacts to SiC with a Stable Protective Coating for Harsh Environment Applications," *ECS Journal of Solid State Science and Technology,* vol. 1, no. 1, pp. P23-P29, 2012.

[71]   K. Ito, T. Onishi, H. Takeda, K. Kohama, S. Tsukimoto, M. Konno, Y. Suzuki, and M. Murakami, "Simultaneous Formation of Ni/Al Ohmic Contacts to Both n- and p-Type 4H-SiC," *Journal of Electronic Materials,* vol. 37, no. 11, pp. 1674-1680, 2008.

[72]   A. Kakanakova-Georgieva, T. Marinova, O. Noblanc, C. Arnodo, S. Cassette, and C. Brylinski, "Characterization of ohmic and Schottky contacts on SiC," *Thin Solid Films,* vol. 343-344, pp. 637-641, 1999.

[73]   S. J. Yang, C. K. Kim, I. H. Noh, S. W. Jang, K. H. Jung, and N. I. Cho, "Study of Co- and Ni-based ohmic contacts to n-type 4H-SiC," *Diamond and Related Materials,* vol. 13, no. 4-8, pp. 1149-1153, 2004.

[74]   J.-C. Cheng, and B.-Y. Tsui, "Reduction of Specific Contact Resistance on n-Type Implanted 4H-SiC Through Argon Inductively Coupled Plasma Treatment and Post-Metal Deposition Annealing," *IEEE Electron Device Letters,* vol. 38, no. 12, pp. 1700-1703, 2017.

[75]   S. Liu, Z. He, L. Zheng, B. Liu, F. Zhang, L. Dong, L. Tian, Z. Shen, J. Wang, Y. Huang, Z. Fan, X. Liu, G. Yan, W. Zhao, L. Wang, G. Sun, F. Yang, and Y. Zeng,

"The thermal stability study and improvement of 4H-SiC ohmic contact," *Applied Physics Letters,* vol. 105, no. 12, pp. 122106, 2014.

[76] A. V. Adedeji, A. C. Ahyi, J. R. Williams, M. J. Bozack, S. E. Mohney, B. Liu, and J. D. Scofield, "Composite Ohmic Contacts to SiC," *Materials Science Forum,* vol. 527-529, pp. 879-882, 2006.
https://doi.org/10.4028/www.scientific.net/MSF.527-529.879

[77] E. D. Luckowski, J. M. Delucca, J. R. Williams, S. E. Mohney, M. J. Bozack, T. Isaacs-Smith, and J. Crofton, "Improved ohmic contact to n-type 4H and 6H-SiC using nichrome," *Journal of Electronic Materials,* vol. 27, no. 4, pp. 330-334, 1998.

[78] R. S. Okojie, D. Lukco, Y. L. Chen, and D. J. Spry, "Reliability assessment of $Ti/TaSi_2/Pt$ ohmic contacts on SiC after 1000 h at 600 C," *Journal of Applied Physics,* vol. 91, no. 10, pp. 6553-6559, 2002. https://doi.org/10.1063/1.1470255

[79] H. Tamaso, S. Yamada, H. Kitabayashi, and T. Horii, "Ti/Al/Si Ohmic Contacts for both n-Type and p-Type 4H-SiC," *Materials Science Forum,* vol. 778-780, pp. 669-672, 2014. https://doi.org/10.4028/www.scientific.net/MSF.778-780.669

[80] S.-J. Joo, S. Baek, S.-C. Kim, and J.-S. Lee, "Simultaneous Formation of Ohmic Contacts on p^+- and n^+-4H-SiC Using a Ti/Ni Bilayer," *Journal of Electronic Materials,* vol. 42, no. 10, pp. 2897-2904, 2013.

[81] T. Ohyanagi, Y. Onose, and A. Watanabe, "Ti/Ni bilayer Ohmic contact on 4H-SiC," *Journal of Vacuum Science & Technology B: Microelectronics and Nanometer Structures,* vol. 26, no. 4, pp. 1359, 2008.
https://doi.org/10.1116/1.2949116

[82] K. Buchholt, R. Ghandi, M. Domeij, C. M. Zetterling, J. Lu, P. Eklund, L. Hultman, and A. L. Spetz, "Ohmic contact properties of magnetron sputtered Ti_3SiC_2 on n- and p-type 4H-silicon carbide," *Applied Physics Letters,* vol. 98, no. 4, pp. 042108, 2011.

[83] H. S. Lee, M. Domeij, C. M. Zetterling, M. Östling, and J. Lu, "Investigation of TiW Contacts to 4H-SiC Bipolar Junction Devices," *Materials Science Forum,* vol. 527-529, pp. 887-890, 2006.
https://doi.org/10.4028/www.scientific.net/MSF.527-529.887

[84] S. K. Lee, S. M. Koo, C. M. Zetterling, and M. Östling, "Ohmic contact formation on inductively coupled plasma etched 4H-silicon carbide," *Journal of Electronic Materials,* vol. 31, no. 5, pp. 340-345, 2002.

[85] N. Kiritani, M. Hoshi, S. Tanimoto, K. Adachi, S. Nishizawa, T. Yatsuo, H. Okushi, and K. Arai, "Single Material Ohmic Contacts Simultaneously Formed on the Source/P-Well/Gate of 4H-SiC Vertical MOSFETs," *Materials Science Forum,* vol. 433-436, pp. 669-672, 2003. https://doi.org/10.4028/www.scientific.net/MSF.433-436.669

[86] S.-C. Chang, S.-J. Wang, K.-M. Uang, and B.-W. Liou, "Investigation of Au/Ti/Al ohmic contact to N-type 4H–SiC," *Solid-State Electronics,* vol. 49, no. 12, pp. 1937-1941, 2005.

[87] S. Tsukimoto, T. Sakai, T. Onishi, K. Ito, and M. Murakami, "Simultaneous formation of p- and n-type ohmic contacts to 4H-SiC using the ternary Ni/Ti/Al system," *Journal of Electronic Materials,* vol. 34, no. 10, pp. 1310-1312, 2005.

[88] T. N. Oder, J. R. Williams, K. W. Bryant, M. J. Bozack, and J. Crofton, "Low Resistance Ohmic Contacts to n-SiC Using Niobium," *Materials Science Forum,* vol. 338-342, pp. 997-1000, 2000. https://doi.org/10.4028/www.scientific.net/MSF.338-342.997

[89] L. J. Evans, R. S. Okojie, and D. Lukco, "Development of an Extreme High Temperature n-Type Ohmic Contact to Silicon Carbide," *Materials Science Forum,* vol. 717-720, pp. 841-844, 2012. https://doi.org/10.4028/www.scientific.net/MSF.717-720.841

[90] F. Roccaforte, F. La Via, V. Raineri, L. Calcagno, and P. Musumeci, "Improvement of high temperature stability of nickel contacts on n-type 6H–SiC," *Applied Surface Science,* vol. 184, no. 1-4, pp. 295-298, 2001.

[91] J. Crofton, P. G. McMullin, J. R. Williams, and M. J. Bozack, "High-temperature ohmic contact to n-type 6H-SiC using nickel," *Journal of Applied Physics,* vol. 77, no. 3, pp. 1317-1319, 1995.

[92] M. G. Rastegaeva, A.N. Andreev, V.V. Zelenin, A.I. Babanin, I.P. Nikitina, V.E. Chelnokov, and V.P. Rastegaev, *Inst. Phys. Conf. ,* vol. 142, pp. 581, 1996.

[93] M. G. Rastegaeva, A. N. Andreev, A. A. Petrov, A. I. Babanin, M. A. Yagovkina, and I. P. Nikitina, "The influence of temperature treatment on the formation of Ni-based Schottky diodes and ohmic contacts to n-6H-SiC," *Materials Science and Engineering: B,* vol. 46, no. 1-3, pp. 254-258, 1997.

[94] T. Uemoto, "Reduction of Ohmic Contact Resistance on n-Type 6H-SiC by Heavy Doping," *Japanese Journal of Applied Physics,* vol. 34, no. Part 2, No. 1A, pp. L7-L9, 1995.

[95] T. Marinova, V. Krastev, C. Hallin, R. Yakimova, and E. Janzén, "Interface chemistry and electric characterisation of nickel metallisation on 6H-SiC," *Applied Surface Science,* vol. 99, no. 2, pp. 119-125, 1996.

[96] E. Kurimoto, H. Harima, T. Toda, M. Sawada, M. Iwami, and S. Nakashima, "Raman study on the Ni/SiC interface reaction," *Journal of Applied Physics,* vol. 91, no. 12, pp. 10215-10217, 2002. https://doi.org/10.1063/1.1473226

[97] F. La Via, F. Roccaforte, V. Raineri, M. Mauceri, A. Ruggiero, P. Musumeci, L. Calcagno, A. Castaldini, and A. Cavallini, "Schottky–ohmic transition in nickel silicide/SiC-4H system: is it really a solved problem?," *Microelectronic Engineering,* vol. 70, no. 2-4, pp. 519-523, 2003.

[98] A. Virshup, F. Liu, D. Lukco, K. Buchholt, A. L. Spetz, and L. M. Porter, "Improved Thermal Stability Observed in Ni-Based Ohmic Contacts to n-Type SiC for High-Temperature Applications," *Journal of Electronic Materials,* vol. 40, no. 4, pp. 400-405, 2010.

[99] C. Deeb, and A. H. Heuer, "A low-temperature route to thermodynamically stable ohmic contacts to n-type 6H-SiC," *Applied Physics Letters,* vol. 84, no. 7, pp. 1117-1119, 2004.

[100] F. La Via, F. Roccaforte, A. Makhtari, V. Raineri, P. Musumeci, and L. Calcagno, "Structural and electrical characterisation of titanium and nickel silicide contacts on silicon carbide," *Microelectronic Engineering,* vol. 60, no. 1-2, pp. 269-282, 2002.

[101] T.-Y. Zhou, X.-C. Liu, C.-C. Dai, W. Huang, S.-Y. Zhuo, and E.-W. Shi, "Effect of graphite related interfacial microstructure created by high temperature annealing on the contact properties of Ni/Ti/6H-SiC," *Materials Science and Engineering: B,* vol. 188, pp. 59-65, 2014.

[102] A. K. Chaddha, J. D. Parsons, and G. B. Kruaval, "Thermally stable, low specific resistance (1.30×10^{-5} $\Omega\,cm^2$) TiC Ohmic contacts to n-type 6H-SiC," *Applied Physics Letters,* vol. 66, no. 6, pp. 760-762, 1995.

[103] U. Schmid, R. Getto, S. T. Sheppard, and W. Wondrak, "Temperature behavior of specific contact resistance and resistivity on nitrogen implanted 6H-SiC with titanium silicide ohmic contacts," *Journal of Applied Physics,* vol. 85, no. 5, pp. 2681-2686, 1999. https://doi.org/10.1063/1.369628

[104] H. Yang, T. H. Peng, W. J. Wang, D. F. Zhang, and X. L. Chen, "Ta/Ni/Ta multilayered ohmic contacts on n-type SiC," *Applied Surface Science,* vol. 254, no. 2, pp. 527-531, 2007.

[105] G. Y. McDaniel, S. T. Fenstermaker, W. V. Lampert, and P. H. Holloway, "Rhenium ohmic contacts on 6H-SiC," *Journal of Applied Physics,* vol. 96, no. 9, pp. 5357-5364, 2004.

[106] K. Gottfried, J. Kriz, J. Leibelt, C. Kaufmann, and T. Gessner, "High temperature stable metallization schemes for SiC-technology operating in air," in 1998 High-Temperature Electronic Materials, Devices and Sensors Conference (Cat. No.98EX132). https://doi.org/10.1109/HTEMDS.1998.730691

[107] T. Jang, L. M. Porter, G. W. M. Rutsch, and B. Odekirk, "Tantalum carbide ohmic contacts ton-type silicon carbide," *Applied Physics Letters,* vol. 75, no. 25, pp. 3956-3958, 1999.

[108] T. Jang, B. Odekirk, L. D. Madsen, and L. M. Porter, "Thermal stability and contact degradation mechanisms of TaC ohmic contacts with W/WC overlayers ton-type 6H SiC," *Journal of Applied Physics,* vol. 90, no. 9, pp. 4555-4559, 2001.

[109] J. Crofton, E. D. Luckowski, J. R. Williams, T. Isaacs-Smith, M. J. Bozack, and R. Siergiej, "Specific contact resistance as a function of doping for n-type 4H and 6H-SiC," *Inst. Phys. Conf. Ser. No 142*, pp. 569-572, 1996.

[110] M. Levit, I. Grimberg, and B. Z. Weiss, "Interaction of Ni90Ti10 alloy thin film with 6H-SiC single crystal," *Journal of Applied Physics,* vol. 80, no. 1, pp. 167-173, 1996.

[111] S. Ferrero, A. Albonico, U. M. Meotto, G. Rambolà, S. Porro, F. Giorgis, D. Perrone, L. Scaltrito, E. Bontempi, L. E. Depero, G. Richieri, and L. Merlin, "Phase Formation at Rapid Thermal Annealing of Nickel Contacts on C-Face n-Type 4H-SiC," *Materials Science Forum,* vol. 483-485, pp. 733-736, 2005. https://doi.org/10.4028/www.scientific.net/MSF.483-485.733

[112] F. Roccaforte, F. La Via, V. Raineri, R. Pierobon, and E. Zanoni, "Richardson's constant in inhomogeneous silicon carbide Schottky contacts," *Journal of Applied Physics,* vol. 93, no. 11, pp. 9137-9144, 2003.

[113] I. Nikitina, K. Vassilevski, A. Horsfall, N. Wright, A. G. O'Neill, S. K. Ray, K. Zekentes, and C. M. Johnson, "Phase composition and electrical characteristics of nickel silicide Schottky contacts formed on 4H-SiC," *Semiconductor Science and*

Technology, vol. 24, no. 5, pp. 055006, 2009. https://doi.org/10.1088/0268-1242/24/5/055006

[114] I. P. Nikitina, K. V. Vassilevski, A. B. Horsfall, N. G. Wright, A. G. O'Neill, C. M. Johnson, T. Yamamoto, and R. K. Malhan, "Structural pattern formation in titanium-nickel contacts on silicon carbide following high-temperature annealing," *Semiconductor Science and Technology,* vol. 21, no. 7, pp. 898-905, 2006. https://doi.org/10.1088/0268-1242/21/7/013

[115] M. R. Rijnders, A. A. Kodentsov, J. A. van Beek, J. van den Akker, and F. J. J. van Loo, "Pattern formation in Pt-SiC diffusion couples," *Solid State Ionics,* vol. 95, no. 1, pp. 51-59, 1997.

[116] J. J. Lander, H. E. Kern, and A. L. Beach, "Solubility and Diffusion Coefficient of Carbon in Nickel: Reaction Rates of Nickel-Carbon Alloys with Barium Oxide," *Journal of Applied Physics,* vol. 23, no. 12, pp. 1305-1309, 1952. https://doi.org/10.1063/1.1702064

[117] A. Hähnel, V. Ischenko, and J. Woltersdorf, "Oriented growth of silicide and carbon in SiC-based sandwich structures with nickel," *Materials Chemistry and Physics,* vol. 110, no. 2-3, pp. 303-310, 2008. https://doi.org/10.1016/j.matchemphys.2008.02.009

[118] C. Y. Kang, L. L. Fan, S. Chen, Z. L. Liu, P. S. Xu, and C. W. Zou, "Few-layer graphene growth on 6H-SiC(0001) surface at low temperature via Ni-silicidation reactions," *Applied Physics Letters,* vol. 100, no. 25, pp. 251604, 2012.

[119] T. Yoneda, M. Shibuya, K. Mitsuhara, A. Visikovskiy, Y. Hoshino, and Y. Kido, "Graphene on SiC(0001) and SiC(000$\bar{1}$) surfaces grown via Ni-silicidation reactions," *Surface Science,* vol. 604, no. 17-18, pp. 1509-1515, 2010.

[120] E. Escobedo-Cousin, K. Vassilevski, T. Hopf, N. Wright, A. O'Neill, A. Horsfall, J. Goss, and P. Cumpson, "Local solid phase growth of few-layer graphene on silicon carbide from nickel silicide supersaturated with carbon," *Journal of Applied Physics,* vol. 113, no. 11, pp. 114309-11, 2013. https://doi.org/10.1063/1.4795501

[121] M. W. Cole, P. C. Joshi, and M. Ervin, "Fabrication and characterization of pulse laser deposited Ni2Si Ohmic contacts on n-SiC for high power and high temperature device applications," *Journal of Applied Physics,* vol. 89, no. 8, pp. 4413-4416, 2001.

Materials Research Forum LLC
doi: http://dx.doi.org/10.21741/9781945291852

[122] W. Lu, W. C. Mitchel, G. R. Landis, T. R. Crenshaw, and W. E. Collins, "Ohmic contact properties of Ni/C film on 4H-SiC," *Solid-State Electronics,* vol. 47, no. 11, pp. 2001-2010, 2003.

[123] M. H. Ervin, K. A. Jones, U. C. Lee, T. Das, and M. C. Wood, "An Approach to Improving the Morphology and Reliability of n-SiC Ohmic Contacts to SiC Using Second-Metal Contacts," *Materials Science Forum,* vol. 527-529, pp. 859-862, 2006. https://doi.org/10.4028/www.scientific.net/MSF.527-529.859

[124] M. H. Ervin, K. A. Jones, U. Lee, and M. C. Wood, "Approach to optimizing n-SiC Ohmic contacts by replacing the original contacts with a second metal," *Journal of Vacuum Science & Technology B: Microelectronics and Nanometer Structures,* vol. 24, no. 3, pp. 1185, 2006. https://doi.org/10.1116/1.2190663

[125] S. Cichoň, P. Macháč, and J. Vojtík, "Ni, NiSi$_2$ and Si Secondary Ohmic Contacts on SiC with High Thermal Stability," *Materials Science Forum,* vol. 740-742, pp. 797-800, 2013. https://doi.org/10.4028/www.scientific.net/MSF.740-742.797

[126] T. Nakamura, and M. Satoh, "Schottky barrier height of a new ohmic contact NiSi$_2$ to n-type 6H-SiC," *Solid-State Electronics,* vol. 46, no. 12, pp. 2063-2067, 2002.

[127] C. Lavoie, C. Detavernier, and P. Besser, "Nickel silicide technology," *Silicide technology for integrated circuits,* L. I. Chen, ed., pp. 95-152, London: the IEE, 2004. https://doi.org/10.1049/PBEP005E_ch5

[128] K. Zekentes, A. Stavrinidis, G. Konstantinidis, M. Kayambaki, K. Vamvoukakis, E. Vassakis, K. Vassilevski, A. B. Horsfall, N. G. Wright, P. Brosselard, S. Q. Niu, M. Lazar, D. Planson, D. Tournier, N. Camara, and M. Bucher, "4H-SiC VJFETs with Self-Aligned Contacts," *Materials Science Forum,* vol. 821-823, pp. 793-796, 2015. https://doi.org/10.4028/www.scientific.net/MSF.821-823.793

[129] K. V. Vassilevski, N. G. Wright, I. P. Nikitina, A. B. Horsfall, A. G. O'Neill, M. J. Uren, K. P. Hilton, A. G. Masterton, A. J. Hydes, and C. M. Johnson, "Protection of selectively implanted and patterned silicon carbide surfaces with graphite capping layer during post-implantation annealing," *Semiconductor Science and Technology,* vol. 20, no. 3, pp. 271, 2005. https://doi.org/10.1088/0268-1242/20/3/003

[130] J. Senzaki, K. Fukuda, and K. Arai, "Influences of postimplantation annealing conditions on resistance lowering in high-phosphorus-implanted 4H–SiC," *Journal of Applied Physics,* vol. 94, no. 5, pp. 2942-2947, 2003. https://doi.org/10.1063/1.1597975

[131] J. C. Cheng, and B. Y. Tsui, "Reduction of Specific Contact Resistance on n-Type Implanted 4H-SiC Through Argon Inductively Coupled Plasma Treatment and Post-Metal Deposition Annealing," *IEEE Electron Device Letters,* vol. 38, no. 12, pp. 1700-1703, 2017. https://doi.org/10.1109/LED.2017.2760884

[132] B. J. Johnson, and M. A. Capano, "Mechanism of ohmic behavior of Al/Ti contacts top-type 4H-SiC after annealing," *Journal of Applied Physics,* vol. 95, no. 10, pp. 5616-5620, 2004.

[133] K. Tone, and J. H. Zhao, "A comparative study of C plus Al coimplantation and Al implantation in 4Hand 6H-SiC," *IEEE Transactions on Electron Devices,* vol. 46, no. 3, pp. 612-619, 1999.

[134] R. Kakanakov, L. Kassamakova, N. Hristeva, G. Lepoeva, N. I. Kuznetsov, and K. Zekentes, "Reliable Ohmic Contacts to LPE p-Type 4H-SiC for High-Power p-n Diode," *Materials Science Forum,* vol. 389-393, pp. 917-920, 2002. https://doi.org/10.4028/www.scientific.net/MSF.389-393.917

[135] F. A. Mohammad, Y. Cao, K. C. Chang, and L. M. Porter, "Comparison of Pt-Based Ohmic Contacts with Ti–Al Ohmic Contacts forp-Type SiC," *Japanese Journal of Applied Physics,* vol. 44, no. 8, pp. 5933-5938, 2005.

[136] N. A. Papanicolaou, A. Edwards, M. V. Rao, and W. T. Anderson, "Si/Pt Ohmic contacts to p-type 4H–SiC," *Applied Physics Letters,* vol. 73, no. 14, pp. 2009-2011, 1998.

[137] S. Tsukimoto, K. Nitta, T. Sakai, M. Moriyama, and M. Murakami, "Correlation between the electrical properties and the interfacial microstructures of TiAl-based ohmic contacts to p-type 4H-SiC," *Journal of Electronic Materials,* vol. 33, no. 5, pp. 460-466, 2004.

[138] R. Kakanakov, L. Kasamakova-Kolaklieva, N. Hristeva, G. Lepoeva, J. B. Gomes, I. Avramova, and T. Marinova, "High Temperature and High Power Stability Investigation of Al-Based Ohmic Contacts to p-Type 4H-SiC," *Materials Science Forum,* vol. 457-460, pp. 877-880, 2004. https://doi.org/10.4028/www.scientific.net/MSF.457-460.877

[139] K. V. Vasilevskii, S. V. Rendakova, I. P. Nikitina, A. I. Babanin, A. N. Andreev, and K. Zekentes, "Electrical characteristics and structural properties of ohmic contacts to p-type 4H-SiC epitaxial layers," *Semiconductors,* vol. 33, no. 11, pp. 1206-1211, 1999.

[140] K. Vassilevski, K. Zekentes, K. Tsagaraki, G. Constantinidis, and I. Nikitina, "Phase formation at rapid thermal annealing of Al/Ti/Ni ohmic contacts on 4H-SiC," *Materials Science and Engineering: B*, vol. 80, no. 1-3, pp. 370-373, 2001.

[141] R. Kakanakov, L. Kassamakova, I. Kassamakov, K. Zekentes, and N. Kuznetsov, "Improved Al/Si ohmic contacts to p-type 4H-SiC," *Materials Science and Engineering: B*, vol. 80, no. 1-3, pp. 374-377, 2001.

[142] J. Crofton, S. E. Mohney, J. R. Williams, and T. Isaacs-Smith, "Finding the optimum Al–Ti alloy composition for use as an ohmic contact to p-type SiC," *Solid-State Electronics*, vol. 46, no. 1, pp. 109-113, 2002.

[143] B. P. Downey, S. E. Mohney, T. E. Clark, and J. R. Flemish, "Reliability of aluminum-bearing ohmic contacts to SiC under high current density," *Microelectronics Reliability*, vol. 50, no. 12, pp. 1967-1972, 2010.

[144] A. Frazzetto, F. Giannazzo, R. L. Nigro, V. Raineri, and F. Roccaforte, "Structural and transport properties in alloyed Ti/Al Ohmic contacts formed on p-type Al-implanted 4H-SiC annealed at high temperature," *Journal of Physics D: Applied Physics*, vol. 44, no. 25, pp. 255302, 2011. https://doi.org/10.1088/0022-3727/44/25/255302

[145] Y. D. Tang, H. J. Shen, X. F. Zhang, F. Guo, Y. Bai, Z. Y. Peng, and X. Y. Liu, "Effect of Annealing on the Characteristics of Ti/Al Ohmic Contacts to p-Type 4H-SiC," *Materials Science Forum*, vol. 897, pp. 395-398, 2017. https://doi.org/10.4028/www.scientific.net/MSF.897.395

[146] M. Vivona, G. Greco, C. Bongiorno, R. Lo Nigro, S. Scalese, and F. Roccaforte, "Electrical and structural properties of surfaces and interfaces in Ti/Al/Ni Ohmic contacts to p-type implanted 4H-SiC," *Applied Surface Science*, vol. 420, pp. 331-335, 2017.

[147] O. Nakatsuka, T. Takei, Y. Koide, and M. Murakami, "Low Resistance TiAl Ohmic Contacts with Multi-Layered Structure for p-Type 4H-SiC," *MATERIALS TRANSACTIONS*, vol. 43, no. 7, pp. 1684-1688, 2002. https://doi.org/10.2320/matertrans.43.1684

[148] M. Vivona, G. Greco, R. L. Nigro, C. Bongiorno, and F. Roccaforte, "Ti/Al/W Ohmic contacts to p-type implanted 4H-SiC," *Journal of Applied Physics*, vol. 118, no. 3, pp. 035705, 2015. https://doi.org/10.1063/1.4927271

[149] S. H. Wang, O. Arnold, C. M. Eichfeld, S. E. Mohney, A. V. Adedeji, and J. R. Williams, "Tantalum-Ruthenium Diffusion Barriers for Contacts to SiC,"

Materials Science Forum, vol. 527-529, pp. 883-886, 2006.
https://doi.org/10.4028/www.scientific.net/MSF.527-529.883

[150] O. Nakatsuka, Y. Koide, and M. Murakami, "CoAl Ohmic Contact Materials with Improved Surface Morphology for p-Type 4H-SiC," *Materials Science Forum,* vol. 389-393, pp. 885-888, 2002. https://doi.org/10.4028/www.scientific.net/MSF.389-393.885

[151] T. Sakai, K. Nitta, S. Tsukimoto, M. Moriyama, and M. Murakami, "Ternary TiAlGe ohmic contacts for p-type 4H-SiC," *Journal of Applied Physics,* vol. 95, no. 4, pp. 2187-2189, 2004.

[152] B. H. Tsao, J. Lawson, and J. D. Scofield, "Ti/AlNi/W and Ti/Ni$_2$Si/W Ohmic Contacts to P-Type SiC," *Materials Science Forum,* vol. 527-529, pp. 903-906, 2006. https://doi.org/10.4028/www.scientific.net/MSF.527-529.903

[153] B. P. Downey, J. R. Flemish, B. Z. Liu, T. E. Clark, and S. E. Mohney, "Current-Induced Degradation of Nickel Ohmic Contacts to SiC," *Journal of Electronic Materials,* vol. 38, no. 4, pp. 563-568, 2009.

[154] R. Konishi, R. Yasukochi, O. Nakatsuka, Y. Koide, M. Moriyama, and M. Murakami, "Development of Ni/Al and Ni/Ti/Al ohmic contact materials for p-type 4H-SiC," *Materials Science and Engineering: B,* vol. 98, no. 3, pp. 286-293, 2003.

[155] B. P. Downey, S. E. Mohney, and J. R. Flemish, "Improved Stability of Pd/Ti Contacts to p-Type SiC Under Continuous DC and Pulsed DC Current Stress," *Journal of Electronic Materials,* vol. 40, no. 4, pp. 406-412, 2010.

[156] L. Kassamakova, R. Kakanakov, N. Nordell, S. Savage, A. Kakanakova-Georgieva, and T. Marinova, "Study of the electrical, thermal and chemical properties of Pd ohmic contacts to p-type 4H-SiC: dependence on annealing conditions," *Materials Science and Engineering: B,* vol. 61-62, pp. 291-295, 1999/07, 1999.

[157] L. Kolaklieva, R. Kakanakov, T. Marinova, and G. Lepoeva, "Effect of the Metal Composition on the Electrical and Thermal Properties of Au/Pd/Ti/Pd Contacts to p-Type SiC," *Materials Science Forum,* vol. 483-485, pp. 749-752, 2005. https://doi.org/10.4028/www.scientific.net/MSF.483-485.749

[158] L. Kassamakova, R. Kakanakov, N. Nordell, and S. Savage, "Thermostable Ohmic Contacts on p-Type SiC," *Materials Science Forum,* vol. 264-268, pp. 787-790, 1998. https://doi.org/10.4028/www.scientific.net/MSF.264-268.787

[159] S. K. Lee, C. M. Zetterling, E. Danielsson, M. Östling, J. P. Palmquist, H. Högberg, and U. Jansson, "Electrical characterization of TiC ohmic contacts to aluminum ion implanted 4H–silicon carbide," *Applied Physics Letters,* vol. 77, no. 10, pp. 1478-1480, 2000.

[160] K. H. Jung, N. I. Cho, J. H. Lee, S. J. Yang, C. K. Kim, B. T. Lee, K. H. Rim, N. K. Kim, and E. D. Kim, "Titanium-Based Ohmic Contact on p-Type 4H-SiC," *Materials Science Forum,* vol. 389-393, pp. 913-916, 2002. https://doi.org/10.4028/www.scientific.net/MSF.389-393.913

[161] J. O. Olowolafe, J. Liu, and R. B. Gregory, "Effect of Si or Al interface layers on the properties of Ta and Mo contacts to p-type SiC," *Journal of Electronic Materials,* vol. 29, no. 3, pp. 391-397, 2000.

[162] B. Pécz, L. Tóth, M. A. di Forte-Poisson, and J. Vacas, "Ti$_3$SiC$_2$ formed in annealed Al/Ti contacts to p-type SiC," *Applied Surface Science,* vol. 206, no. 1-4, pp. 8-11, 2003.

[163] J. Crofton, P. A. Barnes, J. R. Williams, and J. A. Edmond, "Contact resistance measurements on p-type 6H-SiC," *Applied Physics Letters,* vol. 62, no. 4, pp. 384-386, 1993.

[164] F. Moscatelli, A. Scorzoni, A. Poggi, G. C. Cardinali, and R. Nipoti, "Improved electrical characterization of Al–Ti ohmic contacts on p-type ion implanted 6H-SiC," *Semiconductor Science and Technology,* vol. 18, no. 6, pp. 554-559, 2003.

[165] N. Lundberg, and M. Östling, "Thermally stable low ohmic contacts to p-type 6H-SiC using cobalt silicides," *Solid-State Electronics,* vol. 39, no. 11, pp. 1559-1565, 1996.

[166] T. N. Oder, J. R. Williams, M. J. Bozack, V. Iyer, S. E. Mohney, and J. Crofton, "High temperature stability of chromium boride ohmic contacts to p-type 6H-SiC," *Journal of Electronic Materials,* vol. 27, no. 4, pp. 324-329, 1998.

[167] T. Jang, J. W. Erickson, and L. M. Porter, "Effects of Si interlayer conditions on platinum ohmic contacts for p-type silicon carbide," *Journal of Electronic Materials,* vol. 31, no. 5, pp. 506-511, 2002.

[168] J. Crofton, L. Beyer, J. R. Williams, E. D. Luckowski, S. E. Mohney, and J. M. Delucca, "Titanium and aluminum-titanium ohmic contacts to p-type SiC," *Solid-State Electronics,* vol. 41, no. 11, pp. 1725-1729, 1997.

[169] A. A. Iliadis, S. N. Andronescu, K. Edinger, J. H. Orloff, R. D. Vispute, V. Talyansky, R. P. Sharma, T. Venkatesan, M. C. Wood, and K. A. Jones, "Ohmic

contacts to p-6H–SiC using focused ion-beam surface-modification and pulsed laser epitaxial TiN deposition," *Applied Physics Letters,* vol. 73, no. 24, pp. 3545-3547, 1998.

[170] J. S. Shier, "Ohmic Contacts to Silicon Carbide," *Journal of Applied Physics,* vol. 41, no. 2, pp. 771-773, 1970. https://doi.org/10.1063/1.354963

[171] K. V. Vassilevski, V. A. Dmitriev, and A. V. Zorenko, "Silicon carbide diode operating at avalanche breakdown current density of 60 kA/cm^2," *Journal of Applied Physics,* vol. 74, no. 12, pp. 7612-7614, 1993.

[172] W. v. Muench, and I. Pfaffeneder, "Breakdown field in vapor-grown silicon carbidep-njunctions," *Journal of Applied Physics,* vol. 48, no. 11, pp. 4831-4833, 1977.

[173] W. R. King, "Electrical Resistivity of Aluminum Carbide at 990–1240 K," *Journal of The Electrochemical Society,* vol. 132, no. 2, pp. 388, 1985. https://doi.org/10.1149/1.2113847

[174] A. Suzuki, Y. Fujii, H. Saito, Y. Tajima, K. Furukawa, and S. Nakajima, "Effect of the junction interface properties on blue emission of SiC blue LEDs grown by step-controlled CVD," *Journal of Crystal Growth,* vol. 115, no. 1-4, pp. 623-627, 1991.

[175] N. I. Medvedeva, D. L. Novikov, A. L. Ivanovsky, M. V. Kuznetsov, and A. J. Freeman, "Electronic properties of Ti$_3$SiC$_2$-based solid solutions," *Physical Review B,* vol. 58, no. 24, pp. 16042-16050, 1998.

[176] A. Parisini, A. Poggi, and R. Nipoti, "Structural Characterization of Alloyed Al/Ti and Ti Contacts on SiC," *Materials Science Forum,* vol. 457-460, pp. 837-840, 2004. https://doi.org/10.4028/www.scientific.net/MSF.457-460.837

[177] J. L. Freeouf, "Silicide Schottky barriers: An elemental description," *Solid State Communications,* vol. 33, no. 10, pp. 1059-1061, 1980.

[178] F. A. Mohammad, Y. Cao, and L. M. Porter, "Ohmic contacts to silicon carbide determined by changes in the surface," *Applied Physics Letters,* vol. 87, no. 16, pp. 161908, 2005. https://doi.org/10.1063/1.2106005

[179] N. Nordell, S. Savage, and A. Schoner, "Aluminium doped 6H SiC: CVD growth and formation of ohmic contacts," *Inst. Phys. Conf. Ser.,* vol. 142, pp. 573-576, 1995.

[180] K. Vasilevskii, S. Rendakova, I. Nikitina, A. Babanin, A. Andreev, and K. Zekentes, "Electrical characteristics and structural properties of ohmic contacts to

p-type 4H-SiC epitaxial layers," *Semiconductors,* vol. 33, no. 11, pp. 1206-1211, 1999. https://doi.org/10.1134/1.1187850

[181] K. V. Vassilevski, G. Constantinidis, N. Papanicolaou, N. Martin, and K. Zekentes, "Study of annealing conditions on the formation of ohmic contacts on p+4H-SiC layers grown by CVD and LPE," *Materials Science and Engineering B-Solid State Materials for Advanced Technology,* vol. 61-2, pp. 296-300, Jul, 1999. https://doi.org/10.1016/S0921-5107(98)00521-2

[182] T. Troffer, M. Schadt, T. Frank, H. Itoh, G. Pensl, J. Heindl, H. P. Strunk, and M. Maier, "Doping of SiC by Implantation of Boron and Aluminum," *physica status solidi (a),* vol. 162, no. 1, pp. 277-298, 1997.

[183] T. Kimoto, A. Itoh, N. Inoue, O. Takemura, T. Yamamoto, T. Nakajima, and H. Matsunami, "Conductivity Control of SiC by In-Situ Doping and Ion Implantation," *Materials Science Forum,* vol. 264-268, pp. 675-680, 1998. https://doi.org/10.4028/www.scientific.net/MSF.264-268.675

[184] S. K. Roy, K. Vassilevski, N. G. Wright, and A. B. Horsfall, "Silicon Nitride Encapsulation to Preserve Ohmic Contacts Characteristics in High Temperature, Oxygen Rich Environments," *Materials Science Forum,* vol. 821-823, pp. 420-423, 2015. https://doi.org/10.4028/www.scientific.net/MSF.821-823.420

[185] F. Arith, J. Urresti, K. Vasilevskiy, S. Olsen, N. Wright, and A. O'Neill, "Increased Mobility in Enhancement Mode 4H-SiC MOSFET Using a Thin SiO_2/Al_2O_3 Gate Stack," *IEEE Electron Device Letters,* vol. 39, no. 4, pp. 564-567, 2018. https://doi.org/10.1109/LED.2018.2807620

[186] R. Malhan, Y. Takeuchi, I. Nikitina, K. Vassilevski, N. Wright, and A. Horsfall, "Method of forming an ohmic contact in wide band semiconductor," *US patent,* 7,141,498, November 28, 2006.

[187] R. Rupp, R. Kern, and R. Gerlach, "Laser backside contact annealing of SiC power devices: A prerequisite for SiC thin wafer technology," in 25th International Symposium on Power Semiconductor Devices & IC's (ISPSD), 2013, pp. 51-55. https://doi.org/10.1109/ISPSD.2013.6694396

[188] K. Nakashima, O. Eryu, S. Ukai, K. Yoshida, and M. Watanabe, "Improved Ohmic Contacts to 6H-SiC by Pulsed Laser Processing," *Materials Science Forum,* vol. 338-342, pp. 1005-1008, 2000. https://doi.org/10.4028/www.scientific.net/MSF.338-342.1005

[189] Y. Ota, Y. Ikeda, and M. Kitabatake, "Laser Alloying for Ohmic Contacts on SiC at Room Temperature," *Materials Science Forum,* vol. 264-268, pp. 783-786, 1998. https://doi.org/10.4028/www.scientific.net/MSF.264-268.783

[190] R. Rupp, R. Kern, and R. Gerlach, *Production of an integrated circuit including an electrical contact on SiC*, US 8,895,422 B2 2014.

[191] M. de Silva, S. Ishikawa, T. Kikkawa, and S. I. Kuroki, "Low Resistance Ohmic Contact Formation on 4H-SiC C-Face with NbNi Silicidation Using Nanosecond Laser Annealing," *Materials Science Forum,* vol. 858, pp. 549-552, 2016. https://doi.org/10.4028/www.scientific.net/MSF.858.549

[192] M. De Silva, S. Ishikawa, T. Miyazaki, T. Kikkawa, and S.-I. Kuroki, "Formation of amorphous alloys on 4H-SiC with NbNi film using pulsed-laser annealing," *Applied Physics Letters,* vol. 109, no. 1, pp. 012101, 2016.

[193] M. de Silva, T. Kawasaki, T. Kikkawa, and S. I. Kuroki, "Low Resistance Ti-Si-C Ohmic Contacts for 4H-SiC Power Devices Using Laser Annealing," *Materials Science Forum,* vol. 897, pp. 399-402, 2017/05, 2017. https://doi.org/10.4028/www.scientific.net/MSF.897.399

[194] "Information on http://www.wolfspeed.com".

[195] D. P. Hamilton, S. A. Hindmarsh, F. Li, M. R. Jennings, S. A. O. Russell, R. A. McMahon, and P. A. Mawby, "Demonstrating the Instability of SiC Ohmic Contacts and Drain Terminal Metallization Schemes Aged at 300 °C," *Materials Science Forum,* vol. 897, pp. 387-390, 2017. https://doi.org/10.4028/www.scientific.net/MSF.897.387

[196] A. Baeri, V. Raineri, F. Roccaforte, F. La Via, and E. Zanetti, "Study of TiW/Au thin films as metallization stack for high temperature and harsh environment devices on 6H Silicon Carbide," *Materials Science Forum,* vol. 457-460, pp. 873-876, 2004. https://doi.org/10.4028/www.scientific.net/MSF.457-460.873

[197] A. Virshup, L. M. Porter, D. Lukco, K. Buchholt, L. Hultman, and A. L. Spetz, "Investigation of Thermal Stability and Degradation Mechanisms in Ni-Based Ohmic Contacts to n-Type SiC for High-Temperature Gas Sensors," *Journal of Electronic Materials,* vol. 38, no. 4, pp. 569-573, 2009.

[198] R. S. Okojie, D. J. Spry, J. Krotine, C. Salupo, and D. R. Wheeler, "Stable Ti/TaSi$_2$/Pt Ohmic Contacts on N-Type 6H-SiC Epilayer at 600C in Air," *Materials Research Society Symposia Proceedings,* vol. 622, pp. T8.3.1 - 6, 2000.

[199] P. G. Neudeck, D. J. Spry, C. Liang-Yu, G. M. Beheim, R. S. Okojie, C. W. Chang, R. D. Meredith, T. L. Ferrier, L. J. Evans, M. J. Krasowski, and N. F. Prokop, "Stable Electrical Operation of 6H-SiC JFETs and ICs for Thousands of Hours at 500C," *Electron Device Letters, IEEE,* vol. 29, no. 5, pp. 456-459, 2008. https://doi.org/10.1109/LED.2008.919787

[200] P. G. Neudeck, S. L. Garverick, D. J. Spry, L.-Y. Chen, G. M. Beheim, M. J. Krasowski, and M. Mehregany, "Extreme temperature 6H-SiC JFET integrated circuit technology," *physica status solidi (a),* vol. 206, no. 10, pp. 2329-2345, 2009. https://doi.org/10.1002/9783527629077.ch6

[201] P. G. Neudeck, D. J. Spry, L. Y. Chen, R. S. Okojie, G. M. Beheim, R. Meredith, and T. L. Ferrier, "SiC Field Effect Transistor Technology Demonstrating Prolonged Stable Operation at 500 °C," *Materials Science Forum,* vol. 556-557, pp. 831-834, 2007. https://doi.org/10.4028/www.scientific.net/MSF.556-557.831

[202] D. J. Spry, P. G. Neudeck, L. Chen, D. Lukco, C. W. Chang, and G. M. Beheim, "Prolonged 500 C Demonstration of 4H-SiC JFET ICs With Two-Level Interconnect," *IEEE Electron Device Letters,* vol. 37, no. 5, pp. 625-628, 2016. https://doi.org/10.1109/LED.2016.2544700

Materials Research Forum LLC
doi: http://dx.doi.org/10.21741/9781945291852

CHAPTER 3

Schottky Contacts to Silicon Carbide: Physics, Technology and Applications

F. Roccaforte [1*], G. Brezeanu [2], P. M. Gammon [3], F. Giannazzo [1], S. Rascunà [4], M. Saggio [4]

[1] Consiglio Nazionale delle Ricerche – Istituto per la Microelettronica e Microsistemi (CNR-IMM), Strada VIII, 5 – Zona Industriale , 95121 Catania, Italy

[2] University POLITEHNICA of Bucharest, Faculty of Electronics, Telecommunications and Information Technology, 061071 Bucharest, Romania

[3] School of Engineering, University of Warwick, Coventry CV4 7AL, United Kingdom

[4] STMicroelectronics, Stradale Primosole 50, 95121 Catania, Italy

[*] fabrizio.roccaforte@imm.cnr.it

Abstract

Understanding the physics and technology of Schottky contacts to Silicon Carbide is important, for both academic and industrial researchers. In fact, the rectifying contact is a tool for studying carrier transport at metal/semiconductor interfaces, as well as forming the main building block of the Schottky Barrier Diode. In this chapter, the physics of metal/SiC rectifying contacts and the technology of 4H-SiC Schottky diodes are reviewed, presenting a survey of relevant results on this topic, from fundamental concepts of Schottky barriers, to practical information for real device fabrication. Selected examples of 4H-SiC Schottky diodes applications are also briefly discussed.

Keywords

Silicon Carbide, Schottky Contact, Barrier Height, Diode, Wide Band Gap Power Electronics

Contents

1. Introduction

Today, 4H silicon carbide (4H-SiC) is the most mature among the wide band gap semiconductors, and different generation of devices are already commercially available. In particular, owing to the excellent material properties, 4H-SiC devices can allow a significant reduction of the on-state resistance and an increase of the breakdown voltage with respect to their silicon counterparts, resulting in an overall reduction of power losses and more energy efficient components. Hence, SiC-based devices can find applications in many fields of consumer electronics, in automotive and industrial sectors, in conversion systems for renewable energies, transportation, etc.

The first 4H-SiC power device demonstrated was the Schottky Barrier Diode (SBD), whose main building block is the metal/semiconductor Schottky contact. In spite of being relatively simple to fabricate, there are always several physical and technological issues that must be carefully considered to achieve optimal device performance.

In this context, this book chapter aims to provide a concise review on the physics, device technology and applications of Schottky contacts to Silicon Carbide (SiC).

First, the fundamental textbook concepts on the formation of a Schottky barrier are recalled, adapting the description to the specific case of SiC. In particular, the conventional methods to determine the Schottky barrier height by electrical measurements are described. Moreover, an up-to-date survey of Schottky barrier height values are cited from the literature, determined on both n-type and p-type 4H-SiC.

Next, the important fundamental topic of Schottky barrier inhomogeneity in SiC materials is introduced. The effects of contact inhomogeneity on I-V and C-V measurements, and recent results on the modeling of temperature dependent electrical properties of SiC Schottky barriers are reviewed. The characterization methods to probe such inhomogeneity at the nanoscale are also discussed.

Then, a Section is devoted to the technology of the Schottky diode, describing common device layouts and fabrication processes, both for conventional diodes as well as for the Junction Barrier Schottky (JBS) diode. Examples of edge termination for high-voltage diodes are presented. At the end of the Section, the case of the SiC heterojunction diode is discussed as an example of a promising processing step to control the barrier height.

Finally, in the last Section of the chapter, some common applications of SiC Schottky diodes in power electronics and temperature/light sensors technology are explored.

2. Schottky contacts to SiC: fundamentals

The aim of this Section is to introduce the topic of this book chapter, by recalling some fundamental concepts on metal/semiconductor Schottky barriers and reporting an up-to-date survey of literature results on barrier height values measured on both n-type and p-type 4H-SiC.

2.1 Formation of Schottky barrier

It is commonly accepted that metal/semiconductor contacts divide into two distinct categories, i.e., the Ohmic and the Schottky contacts.

Ohmic contacts have linear and symmetric current-voltage characteristics for positive and negative applied bias. Conversely, in a rectifying Schottky contact the current flow is asymmetric, i.e., it is favored under forward bias and inhibited under reverse bias.

The most important parameter that describes a metal/semiconductor contact is the *Schottky Barrier Height* (SBH). To introduce this fundamental physical concept, it is useful to refer to the classical graphical description of the SBH formation.

Advancing Silicon Carbide Electronics Technology I
Materials Research Foundations **37** (2018)

Materials Research Forum LLC
doi: http://dx.doi.org/10.21741/9781945291852

Fig. 1a shows the energy band diagram of a metal and an n-type semiconductor before they are put into intimate contact. The metal and the semiconductor work functions, i.e., $q\Phi_m$ and $q\Phi_s$, are the energies required to bring an electron from the Fermi level of the material into the vacuum. The semiconductor electron affinity $q\chi_s$ is the energy difference between the vacuum level and the bottom of the semiconductor conduction band E_C.

When the metal and the semiconductor are brought into intimate contact, if the semiconductor work function $q\Phi_s$ is lower than the metal work function $q\Phi_m$ (as occurs for most metals on SiC), electrons will flow from the n-type semiconductor into the metal, leaving behind a region of positively charged donors over the depletion width W.

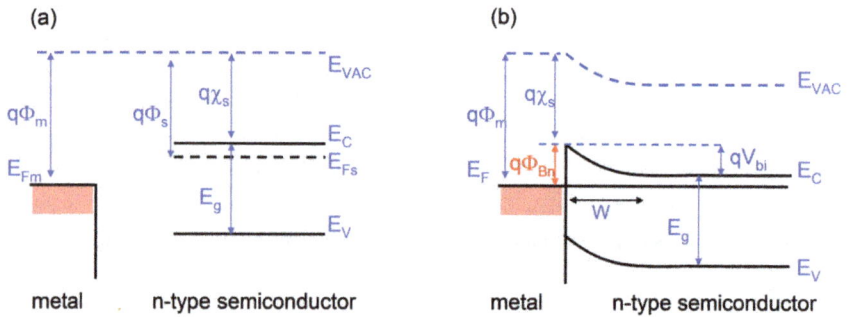

Figure 1. Schematic representation of the energy band diagram for a metal contact onto a n-type semiconductor, before (a) and after (b) they are brought into intimate contact, showing the formation of the Schottky barrier height $q\Phi_{Bn}$.

This charge transfer proceeds until thermodynamic equilibrium is achieved and the two Fermi levels are aligned. In this condition, the energy level of the electrons in the semiconductor will be raised near the metal/semiconductor interface by an amount qV_{bi}, as shown in Fig. 1b. The amount V_{bi} is often referred to as the contact's built-in potential.

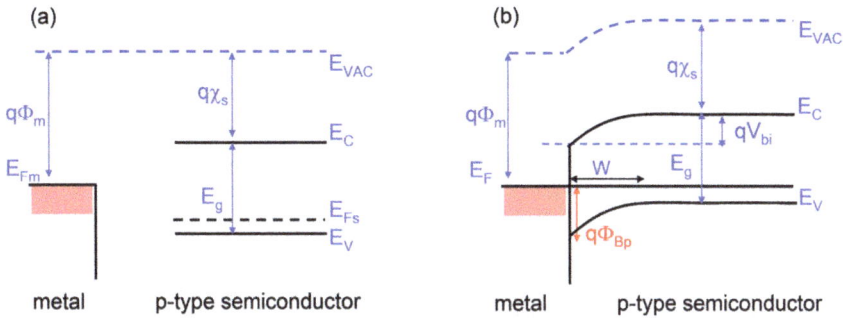

Figure 2. Schematic representation of the energy band diagram for a metal contact onto a p-type semiconductor, before (a) and after (b) they are brought into intimate contact, showing the formation of the Schottky barrier height $q\Phi_{Bp}$.

According to the well-known *Schottky-Mott* relationship, the *Schottky barrier height* $q\Phi_{Bn}$ in an *n*-type material can be defined as the difference between the metal work function $q\Phi_m$ and the semiconductor electron affinity $q\chi_s$ [1]:

$$q\Phi_{Bn} = q(\Phi_m - \chi_s) \tag{1}$$

The Schottky barrier height is a fundamental parameter, which determines the electrical behavior of a metal/semiconductor contact. The Schottky barrier $q\Phi_{Bn}$ can be seen as the energy necessary for electrons in the metal to penetrate into the semiconductor, while qV_{bi} is the barrier seen from the electrons in the semiconductor.

Analogously, the formation process of a Schottky barrier of a metal in contact with a *p*-type semiconductor is schematically shown in Figs. 2a and 2b. In this case, the Schottky barrier height $q\Phi_{Bp}$ is written as:

$$q\Phi_{Bp} = E_g - q(\Phi_m - \chi_s) \tag{2}$$

where E_g is the value of the forbidden energy gap of the semiconductor. Hence, according to the Eqs. 1 and 2, the sum of the SBHs of a metal onto a *n*-type and *p*-type semiconductor should be equal to its band gap, i.e., $q(\Phi_{Bn} - \Phi_{Bp}) = E_g$.

Considering the most common case of an *n*-type semiconductor, it is worth noting that the SBH is almost independent of the donor concentration N_D. Actually, there exists a dependence of Φ_{Bn} on N_D through the image force lowering of the barrier $\Delta\Phi_{Bn}$ ($\Delta\Phi_{Bn} \propto$

$N_D^{1/4}V^{1/4}$) [1]. On the other hand, the barrier width W depends on the reverse of the square root of the doping level, i.e., $W \propto N_D^{-1/2}$.

In general, according to the classical description, the mechanism ruling the current transport at a metal/semiconductor contact depends on the doping level of the semiconductor N. In particular, for lightly doped semiconductors ($N < 1\times10^{17}$ cm^{-3}) the main conduction mechanism is the *thermionic emission* (TE), i.e., the carriers having sufficient thermal energy to surmount the Schottky barrier can pass from one material to the other. For intermediate semiconductor doping levels (N in the range 10^{17}-10^{19} cm^{-3}) the current transport through the barrier will be ruled by the *thermionic field emission* (TFE). This conduction mechanism involves the carriers that do not have sufficient thermal energy to surmount the barrier (as in the case of the TE) but their thermal energy is sufficient to tunnel at an energy higher than the Fermi level, i.e., where the barrier is thinner. Finally, for heavily doped semiconductors ($N > 1\times10^{19}$ cm^{-3}), the small depletion layer W will result in a very thin barrier. Under this condition, the transport through the barrier is ruled by the *field emission* (FE) mechanism, as the carriers can easily tunnel through this thin barrier.

In general, *n*-type SiC is used for the fabrication of Schottky diodes, due to the high mobility of electrons with respect to the holes and to the low SBH values obtained in a *n*-type material with respect to the *p*-type. The typical doping level used for the fabrication of the drift layer in SiC Schottky diodes is in the range of 10^{15}-10^{16} cm^{-3}. Hence, under these conditions thermionic emission (TE) is the typical carrier transport mechanism. In this case, under the application of a voltage V across a metal/semiconductor junction the current I through the contact will be given by [1]:

$$I = AA^*T^2e^{-\frac{q\Phi_B}{kT}}\left(e^{\frac{qV}{nkT}}-1\right) \tag{3}$$

where A is the contact area, A^* is the *Richardson constant*, q is the elementary charge, k is the Boltzmann constant, T the absolute temperature and n is the so called *ideality factor*. This latter parameter incorporates all those effects that make the contact non-ideal (e.g., deviations from the TE transport, dependence of the barrier on the applied voltage, spatial inhomogeneity of the barrier, etc.).

2.2 Experimental determination of the Schottky barrier height

The value of the Schottky barrier height at a metal/SiC contact can be experimentally determined using different methods. The most common ones are electrical methods based on current-voltage (*I-V*, *I-V-T*) or capacitance-voltage (*C-V*) measurements. For this purpose, a simple Schottky diode is typically used as test vehicle for the electrical measurements, as schematically depicted in Fig. 3a.

An epitaxial layer with a doping concentration in the order of 10^{15}-10^{16} cm^{-3} is grown onto a heavily doped substrate (with a doping level in the order of 10^{18}-10^{19} cm^{-3}). The Ohmic contact on the back side is typically fabricated using Nickel Silicide (Ni$_2$Si) obtained by the sintering of Nickel films subjected to thermal annealing processes at high temperatures [2].

When the diode is forward biased, for $V >> kT/q$, the current in Eq. 3 can be written as:

$$I = AA^*T^2 e^{-\frac{q\Phi_B}{kT}} e^{\frac{qV}{nkT}} = I_S e^{\frac{qV}{nkT}} \tag{4}$$

where the so called *saturation current* I_S is given by:

$$I_S = AA^*T^2 e^{-\frac{q\Phi_B}{kT}} \tag{5}$$

By plotting the forward *I-V* characteristics in a semi-logarithmic scale, it is possible to fit Eq. 4 to the experimental data in the linear region. This allows the determination of the saturation current, I_S, by extrapolating this fit to the intercept with the y-axis (at $V = 0$) and, hence, the extraction of barrier height $q\Phi_B$ from Eq. 5. Moreover, from the slope of the linear fit, the ideality factor n can be also determined. The application of this method requires the knowledge of the value of the Richardson constant A^*. However, as A^*

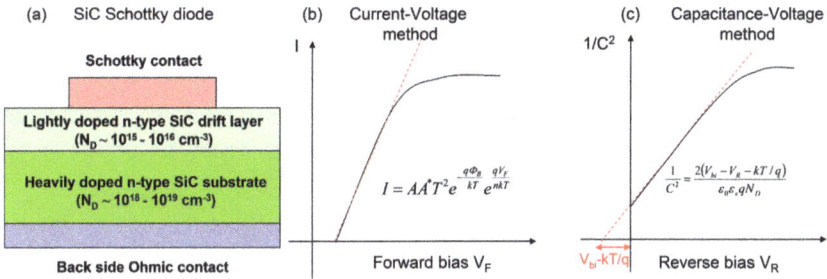

Figure 3. (a) Schematic of a simple SiC Schottky diode used for the determination of the SBH; (b) Extraction of the SBH from the linear fit of the forward I-V characteristic of a Schottky diode in a semi-logarithmic scale; (c) Extraction of the SBH from the linear fit of the plot of $1/C^2$ as a function of the reverse bias V_R.

Advancing Silicon Carbide Electronics Technology I   Materials Research Forum LLC
Materials Research Foundations **37** (2018)   doi: http://dx.doi.org/10.21741/9781945291852

appears in the logarithmic term, only very small errors in the determination of $q\Phi_B$ are done using this method [3].

Similarly, from the *I-V* characteristics acquired at different temperatures (*I-V-T*), a plot of *ln(I_S)* as a function of *1/kT* (termed the *Richardson plot*) enables the determination of both the barrier height $q\Phi_B$ and the Richardson constant A^*. The commonly used value of A^* for 4H-SiC is 146 A·cm^{-2}·K^{-2}.

Here, it is worth noting that considering the series resistance R_s of the diode in Eq. 3, a more complete expression of the current is obtained:

$$I = AA^*T^2 e^{-\frac{q\Phi_B}{kT}}\left(e^{\frac{q(V-IR_s)}{nkT}} - 1\right)$$

(6)

which re-written in terms of the current density J ($= I/A$) becomes:

$$J = A^*T^2 e^{-\frac{q\Phi_B}{kT}}\left(e^{\frac{q(V-JR_{ON})}{nkT}} - 1\right) = J_S\left(e^{\frac{q(V-JR_{ON})}{nkT}} - 1\right)$$

(7)

where J_S is the *saturation current density* and R_{ON} is the so-called specific on-resistance of the devices (measured in Ω·cm^2). These concepts will be used again in Section 4.

The SBH can be also determined from capacitance-voltage (*C-V*) measurements on Schottky diodes. In particular, the *capacitance per unit of area C* of the depletion layer W is given by:

$$C = \sqrt{\frac{\varepsilon_0\varepsilon_s qN_D}{2(V_{bi} - V_R - kT/q)}}$$

(8)

where ε_0 and ε_s are the vacuum and semiconductor permittivity, respectively, N_D is the donor density of the drift layer, V_{bi} is the build-in potential and V_R is the applied reverse bias. Inverting and squaring this equation leads to:

$$\frac{1}{C^2} = \frac{2(V_{bi} - V_R - kT/q)}{\varepsilon_0\varepsilon_s qN_D}$$

(9)

Hence, by plotting $1/C^2$ as a function of the applied reverse bias V_R, the value of V_{bi} can be determined from the intercept of a linear fit of the date with the x-axis. Once V_{bi} is known, the barrier height $q\Phi_{Bn}$ can be extracted from the relation:

$$\Phi_{Bn} = V_{bi} + V_n$$

(10)

where qV_n is the distance of the conduction band edge E_C from the semiconductor Fermi level E_F, i.e., $V_n = [kT/q]ln(N_C/N_D)$ with N_C the effective density of states in the conduction band.

Moreover, the slope of the linear fit of the $1/C^2$ vs V_R plot allows the extraction of the doping level of the epitaxial layer N_D.

Electrical analyses based on I-V and C-V measurements on Schottky diodes are the most used techniques to determine the SBH values of metal contacts to SiC. However, it is worth mentioning that also *photoemission measurements* are sometimes employed to determine the barrier height. However, this optical method requires special equipment and semi-transparent Schottky contacts, thus making it less common with respect to the electrical measurements. Further details can be found in Ref. [3].

2.3   Schottky barriers to *n*-type and *p*-type SiC

In the last two decades, Schottky contacts on SiC have been widely investigated on the most common SiC polytypes, 3C-SiC [4,5,6,7], 6H-SiC [8,9,10,11,12,13,14,15,] and 4H-SiC [16,17,18,19,20,21,22,23,24,25,26,27,28]. In this chapter, the focus is on 4H-SiC, the most common polytype for power electronics applications. Table 1 and Table 2 report experimental values of Schottky barrier height for different metal contacts on *n*-type $(q\Phi_{Bn})$ and *p*-type $(q\Phi_{Bp})$ 4H-SiC, respectively. These values are just a small selection of a significant quantity of data reported in the literature. These have been determined either by current-voltage (I-V) of by capacitance-voltage (C-V) analysis of 4H-SiC Schottky diodes fabricated on the Si-face of the material.

The most commonly used metals for Schottky contacts are Titanium (Ti) and Nickel (Ni), which typically show a higher reproducibility of the barrier height, and have been easily integrated in the fabrication of Schottky diodes. In particular, Ti has a lower barrier (about 1.2 eV) and is currently used as Schottky contact in many commercial diodes. However, its barrier can be extremely sensitive to deposition method and post-deposition thermal budget [26]. For that reason, Molybdenum (Mo) become a recent candidate to be considered as a barrier metal for 4H-SiC Schottky diodes since it is less sensitive to thermal budget and provides relatively low Schottky barrier on 4H-SiC.

Table 1. Schottky Barrier Height ($q\Phi_{Bn}$) for different metals on n-type 4H-SiC (0001). The values were determined by I-V or C-V measurements on Schottky diodes.

SBH values on *n*-type 4H-SiC(0001)				
metal	$q\Phi_{Bn}$ (eV)		*n*	Annealing Temp.	Ref.
	I-V	*C-V*			
Mo	1.04	1.08	1.04	As deposited	[16]
Mo	1.11		1.03	As deposited	[17]
Mo	1.21		1.02	600°C	[17]
W	1.17		1.04	As deposited	[17]
W	1.09		1.04	600°C	[17]
Ta	1.10		1.02	N.A.	[18]
Ti	1.10	1.15	1.03	As deposited	[24]
Ti	1.20		1.23	As deposited	[11]
Ti	1.20	1.21	1.03	400°C	[16]
Ti	1.23	1.32	1.02	500°C	[26]
Ti	1.27		1.04	N.A.	[23]
Ti	1.27			As deposited	[19]
Ti$_{0.58}$W$_{0.42}$	1.22	1.23	1.05	As deposited	[20]
Ti$_{0.58}$W$_{0.42}$	1.18	1.19	1.10	500°C	[20]
Ni	1.32			As deposited	[21]
Ni	1.40		1.1	As deposited	[22]
Ni	1.45	1.65	1.10	As deposited	[16]
Ni	1.60	1.70	1.01	As deposited	[24]
Ni$_2$Si	1.60		1.05	Ni +700°C	[23]
Au	1.73	1.80	1.08	As deposited	[24]
Pt	1.39		1.01	Sputter at 200°C	[25]
Pt	1.817	1.883	1.08	As deposited	[27]
Pd	0.71	1.18	1.35	As deposited	[28]
Pd	0.89	1.30	1.04	300°C	[28]

On the other hand, Ni has the advantage to form silicides (e.g., Ni$_{31}$Si$_{12}$, Ni$_2$Si) after annealing (> 500 °C), resulting in an ideal and reproducible barrier that is almost independent of the surface preparation [29]. However, the high Ni$_2$Si/4H-SiC SBH (about 1.6 eV) [23] makes this contact less useful for power devices, although it can be promising for other applications (e.g., temperature sensors, UV-detectors, etc.) where other peculiarities of that contact, such as a lower leakage and a "self-aligned" process (like the Ni$_2$Si formation), can be beneficial for the realization of semi-transparent interdigitated electrodes [30]. Some example of these applications will be described in Section 5.

Table 2. Schottky Barrier Height ($q\Phi_{Bp}$) for different metals on p-type 4H-SiC (0001). The values were determined by I-V or C-V measurements on Schottky diodes.

metal	$q\Phi_{Bp}$ (eV)		n	Annealing temp.	Ref.
	I-V	*C-V*			
Ti	1.94	2.07	1.07	300°C	[36]
Ti/Al	1.4	1.5	2.2	N.A.	[34]
Ti$_{0.58}$W$_{0.42}$	1.91	1.66	1.08	500°C	[20]
Ni	1.35	1.49	1.08	300°C	[36]
Au	1.31	1.56	1.29	300°C	[36]
Au	1.57			N.A.	[37]

SBH values on p-type 4H-SiC(0001)

Besides the standard solutions shown in Tables 1 and 2, "non conventional" metallization schemes, like metal borides [31], or rare earth oxides [32], have been also investigated, in order to achieve a better thermal stability of the barrier. However, these are not practical for power devices. Recently, *Stöber et al.* [33] explored molybdenum nitride (Mo$_2$N) films as Schottky barriers to 4H-SiC. By varying the nitrogen content during the sputter of the metal film it was possible to tailor the effective Schottky barrier height in the range 0.68–1.03 eV. However, the ideality factor and leakage current of the diodes are still to be optimized.

Typically, the experimental values of $q\Phi_B$ measured on 4H-SiC are larger than those measured in the case of 6H-SiC and 3C-SiC [34,35]. This behavior can be explained by the different electron affinity of the three polytypes, i.e. 3.8 eV in 3C-SiC, 3.3 eV in 6H-SiC, and 3.1 eV in 4H-SiC [35].

Less work has been carried out in the case of *p*-type 4H-SiC [20,36,37,38]. These works are rather old, given that the necessary heavily doped *p*-type substrates are no longer commercially available.

The SBH values are reported in Figs. 4a and 4b as a function of the metal work function $q\Phi_m$ for both *n*-type and *p*-type 4H-SiC. As can be seen, a general trend can be observed, with the values of the barrier in *n*-type material $q\Phi_{Bn}$ that increase with increasing metal work function $q\Phi_m$, and those on *p*-type material $q\Phi_{Bp}$ that decrease with increasing $q\Phi_m$.

The slope of the Φ_B vs Φ_m plot is the so-called *interface index S*, describing the variation of the measured barrier height with respect to the metal work function ($\partial\Phi_B/\partial\Phi_m$). The interface index S gives information on the ideality of the metal/semiconductor Schottky contact. In particular, an interface index $S = 1$ is expected for an ideal contact obeying to the Schottky-Mott relation (Eq. 1). However, in real cases where surface states are present at the metal/semiconductor interface, the Schottky-Mott relation will no longer be valid and the Schottky barrier height $q\Phi_B$ will be more weakly dependent on the metal work function $q\Phi_m$, thus leading to an interface index S < 1. In the extreme case, if the density of surface states is very large, the Fermi level at the surface of the semiconductor will be pinned at a level $q\Phi_0$, and $q\Phi_B$ will be independent of $q\Phi_m$, given by the relation:

$$q\Phi_B = E_g - q\Phi_0 \qquad (11)$$

This is known as the *Bardeen-limit*. In this case, the complete Fermi level pinning occurs and the properties of the interface are independent of the metal ($S = 0$) [39,40].

For the data reported in Figs. 4a and 4b it can be deduced that the interface index S is lower than 1, meaning that a partial Fermi level pinning at the surface occurs.

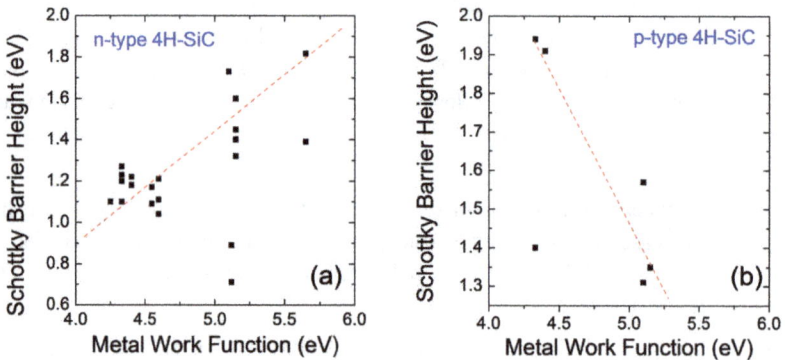

Figure 4. (a) Schottky barrier height $q\Phi_{Bn}$ as a function of the metal work function $q\Phi_m$ for several metals contacts on n-type 4H-SiC; (b) Same plot of $q\Phi_{Bp}$ vs $q\Phi_m$ for metal contacts on p-type 4H-SiC. The data are taken from Tables 1 and 2.

It is worth noting that different values of the interface index are reported in the literature. As an example, *Zhao et al.* [34] reported a wide selection of literature data relative to both 6H- and 4H-SiC, finding interface index values of 0.6 and 0.9, respectively. On the other hand, *Kimoto* [41] compared the behavior of some common metals (Ti, Mo, Ni, Au, Pt) on 4H-SiC epilayers with three different orientations, i.e., (0001), (000$\bar{1}$) and (11$\bar{2}$0). Interestingly, higher values of the SBH were found on the (000$\bar{1}$) with respect to the (0001), while the values of SBH found on (11$\bar{2}$0) were in between. This behavior was ascribed to the different polarity of the faces or to a different distribution of the metal/SiC interface states. In each case, the interface index S was in the range 0.8 – 0.9, suggesting that only a moderate Fermi level pinning occurs and the system is close to the Schottky-Mott limit. In this context, the relevance of the interface states was pointed out by *Roccaforte et al.* [35], who compared the behavior of non-annealed metals on the most common SiC polytypes. In this case, lower values of the interface index S were found for the hexagonal 6H-SiC and 4H-SiC (0.41-0.42), thus demonstrating that annealing of metal/SiC interfaces is beneficial to approach the ideal limit.

According to the model proposed by *Kurtin et al.* [42], the interface index is expected to increase with the degree of ionicity of the semiconductor, defined as the difference in the electronegativities of the two components ($\Delta\chi_s$). Hence, in SiC ($\Delta\chi_s$=0.65 eV), Schottky contacts are expected to give $S < 1$ [42]. Indeed, only a few authors reported that under specific surface treatments an "un-pinned" Fermi level ($S \approx 1$) can be achieved in metal/6H-SiC contacts, with the virtual elimination of interface states and the achievement of the ideal barrier value [43].

Apparently, surface preparation before metal deposition and post-deposition thermal annealings are very important issues, which can have a significant impact on the value of the SBH [24,29,32,44,45,46,47]. The surface roughness, eventual processing-related contaminants at the interface, a residual thin oxide layer, etc. can affect the uniformity of the Schottky barrier and, consequently, lead to non-ideal I–V characteristics of Schottky contacts. As an example, this aspect will be described in Section 3, where the nanoscale homogeneity of the Schottky barrier in the presence of different surface treatments will be described as case study.

2.4   Forward and reverse characteristics of 4H-SiC Schottky diodes

Fig. 5 shows, as an example, the forward and reverse characteristics of 4H-SiC Schottky diodes using Ti and Ni$_2$Si as barrier metal [48]. The characteristics of the diodes, reported in a semi-logarithmic plot, have been acquired at different temperatures.

As can be seen, under forward bias the current shows an almost ideal behavior and a wide linearity range in the semi-logarithmic plot. Moreover, the forward current increases with increasing the measurement temperature, as predicted by the TE theory. For a fixed current, a low voltage drop is measured in the Ti/4H-SiC contacts with respect to the Ni$_2$Si/4H-SiC contacts, due to the low Schottky barrier height of Ti with respect to Ni$_2$Si (see Table 1). The occurrence of two different slopes in the forward characteristics of Ti/4H-SiC Schottky diode is visible at low temperatures, thus being related to the presence of a Schottky barrier inhomogeneity. This latter aspect will be discussed in detail in Section 3.

On the other hand, under reverse bias, a significant increase of the leakage current occurs for high reverse voltage, i.e., where a high electric field produces an enhanced image force lowering effect on the barrier. Clearly, a higher leakage current is measured in the Ti/4H-SiC contacts with respect to the Ni₂Si/4H-SiC. In Section 4 it will be shown that in order to prevent soft reverse characteristics (as those shown in Fig. 5), in high-voltage applications the simple Schottky barrier diodes are replaced by more complex Schottky devices integrating PN junctions in the surface region.

Moreover, it is worth mentioning that in SiC diodes the electric field strength in the depletion region under reverse bias can be about 10 times higher than in Si-diodes. Such a high electric field results in strong band bending and, hence, a thin potential barrier, which can be tunneled through by electrons. Therefore, while in Si diodes the reverse current is typically described by the thermionic emission model accounting the image force lowering of the barrier, the reverse characteristics of 4H-SiC Schottky diodes has been well described using the thermionic field emission (TFE) mechanism [49,50].

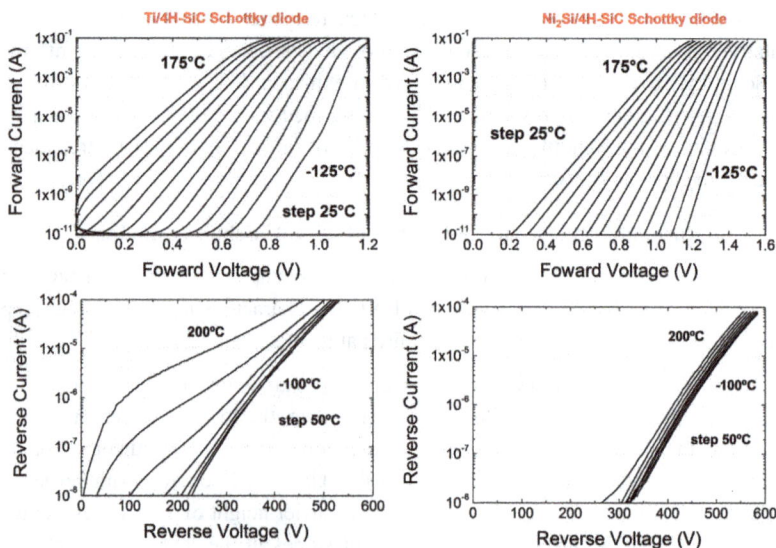

Figure 5. Forward and reverse I-V characteristics of 4H-SiC Schottky diodes using Ti and Ni₂Si as barrier metal, acquired at different temperatures. Reprinted from Ref. [48], with the permission of Trans Tech Publications.

3.   Schottky barrier inhomogeneity in SiC

Section 2 referred to an ideal metal/semiconductor Schottky contact, i.e., a contact with a homogeneous Schottky barrier height (SBH) across the entire area, both on the macro scale (e.g., from edge to center of the diode) and on the nanoscale (e.g., on a scale less than the Debye length). According to this assumption, the Schottky diode parameters are often extracted from the *I-V* characteristics measured at constant temperature using the nominal area. This analysis is the overwhelmingly preferred method, present in most of the literature pertaining the Schottky diode characterization.

A "homogeneous" metal/semiconductor Schottky contact can be defined as one with a single barrier ($q\Phi_{Bn}$) and ideality factor n close to 1. According to the TE mechanism, $q\Phi_{Bn}$ and n should be reasonably temperature and voltage independent. Furthermore, the barrier height extracted will be independent of the measurement technique (*I-V, C-V, I-V-T*, or photocurrent measurements), while the absence of charge at the interface would lead to a barrier height close to that expected from the Schottky-Mott approximation (Eq. 1).

In the early 1980s, several theories into "Schottky contact inhomogeneity" were developed [51,52,53,54,55,56,57], mostly on Silicon, to explain the discrepancies from the TE behavior, the variation of barrier heights and ideality factors with temperature, and barrier heights that tended to be higher when extracted from C-V measurements than from *I-V* techniques.

Later, the rapid maturity of SiC materials, combined with the relative simplicity of producing metal/semiconductor contacts, has led to an impressive increase of publications on metal/SiC Schottky diodes addressing the topic of Schottky barrier inhomogeneity [23,35, 58,59,60,61,62,63,64,65,66,67,68,69,70,71,72,73]. The reason for this recent upsurge was mainly due to a large variety of near-surface defects (basal plane dislocations, polytype inclusions, carrots, growth pits and micropipes) present in SiC and affecting the SBH properties. Moreover, the requirement of 2-8° off-axis epilayers (leaving atomic terracing on the surface), and the presence of undesired processing remnants (such as dirt or surface contamination), also contributed to barrier inhomogeneity. The following paragraphs of this Section describe in more detail the experimental evidence of Schottky barrier inhomogeneity in SiC, and the models and characterization techniques applied to metal/SiC systems.

3.1   The experimental evidence of SBH inhomogeneity

When the forward *I-V* characteristics of an inhomogeneous Schottky diode are modeled with the ideal metal/semiconductor contact theory (Eqs. 3-5) over extended temperature ranges, an anomalous behavior deviating from the pure TE mechanism can be observed. Most common, the presence of large ideality factors is often observed at low temperature, accompanied by a significant temperature or voltage dependence of both SBH $q\Phi_{Bn}$ and ideality factor n [23,56,57,59,61,64,71,73]. Moreover, double turn-on effects (double bumps) [58] are also observed, the forward characteristics appearing as if they were two (or even more) diodes with parallel conduction paths. This in particular, is a sign of some likely macroscopic differentiation in the barrier heights, such that a single $q\Phi_{Bn}$ could not hold true. Also commonly seen is variability of characteristics from chip to chip, or even between like devices on the same chip [58,59,60]. Finally, when comparing the $q\Phi_{Bn}$ extracted from I-V analysis to that from other techniques (e.g., C-V measurements), one will often find that $\Phi_{Bn,(C\text{-}V)} > \Phi_{Bn,(I\text{-}V)}$ [51,54,56,57]. This is the result of the presence of a distribution of SBHs [52,57], with current passing preferentially over the lowest barriers first, making $q\Phi_{Bn,(I\text{-}V)}$ appear low compared to the average in $q\Phi_{Bn,(C\text{-}V)}$ [3].

In the case of metal/SiC Schottky structures, these spurious effects can be attributed to the presence of interfacial inhomogeneity, due to doping non-uniformity, elevated interface state density, surface defects, a mixture of different metal semiconductor phases etc.. Furthermore, because the interface between metal and semiconductor is not atomically flat, the built-in potential and the barrier height are expected to fluctuate spatially.

Fig. 6 illustrates many of these inhomogeneous effects, observed in one case study of Mo/4H-SiC SBDs fabricated on a 35 µm thick n-type epilayer ($N_D = 4\times10^{15}$ cm^{-3}) [66]. The Schottky contacts of these diodes were formed using a 200 nm thick Mo film, annealed at 500 °C.

First, the forward J-V characteristics of the Mo/4H-SiC diode are shown in Fig. 6a over a wide temperature range (40-320 K). At room temperature, a barrier height of 1.36 eV and an ideality factor of 1.021 can be extracted. Fig. 6b shows the semilog plot of the same J-V characteristics taken at three of the temperatures, together with the theoretical TE fits (dotted lines). This plot demonstrates that even at room temperature, there is always some voltage dependence of the ideality factor, which becomes more exaggerated at lower temperatures. As will be described further later, this effect is related to parallel conduction paths occurring through regions of different barrier height. However, plotting the ideality factor and SBH as a function of temperature (Fig. 6c), it is possible to notice that the ideality factor will steadily climb to values far exceeding 2 below 100 K, while the SBH mirrors this by decreasing at reduced temperature. According to prevailing theory [1], such high values of n should suggest the dominance of recombination current transport. Instead, SBH inhomogeneity theory suggests that this is the inherent failure of the ideality factor concept. The positive SBH-temperature relationship is also counterintuitive, given that a negative temperature dependence exists for the bandgap, while any freeze-out of the dopant should also result in a negative SBH-temperature relationship.

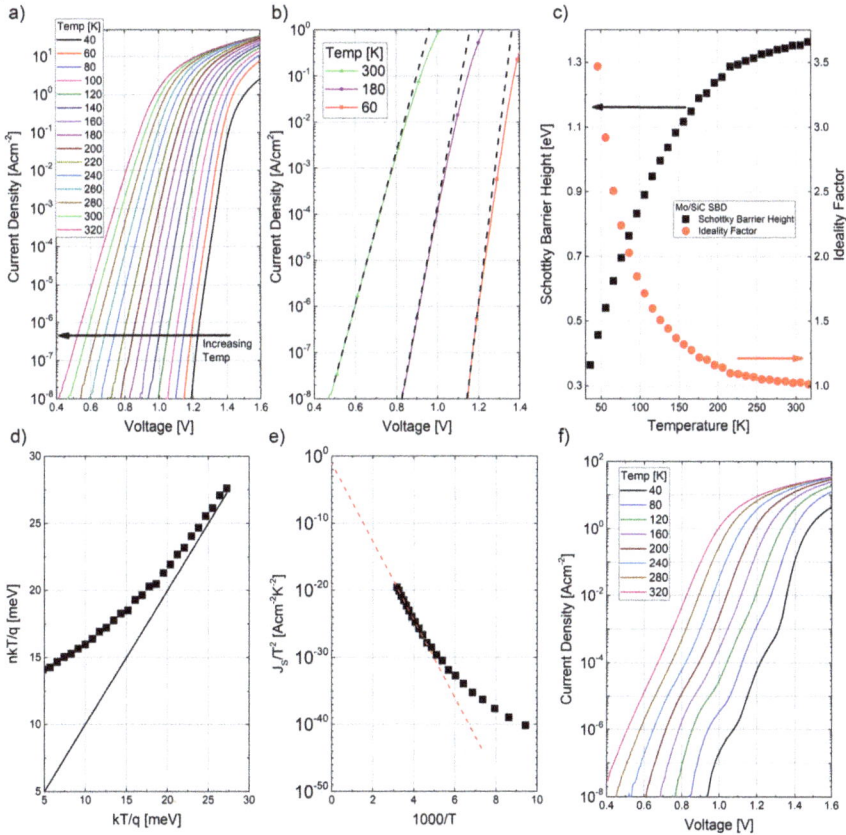

Figure 6. The effects of SBH inhomogeneity shown on a Mo/SiC Schottky diode. (a) J-V-T characteristics of a single device. (b) Zoom on three of the temperatures, showing increasing voltage dependency at low temperature. (c) Extracted $q\Phi_{Bn}$ and n across the temperature range. (d) Plot of nkT/q versus kT/q showing the temperature dependence of the ideality factor. (e) Richardson's plot log(J_S/T^2) versus 1/T. (f) J-V-T characteristics acquired on a second device showing typical double bumps.

One of the first ways to visualize the SBH inhomogeneity was reported in 1969 by *Saxena* [74], who proposed the plot of nkT versus kT to depict the temperature dependence of the ideality factor. Figure 6d shows such a plot in our case study of the Mo/4H-SiC diode. The continuous line in this graph represents the ideal case $n = 1$, while the scattered symbols are the experimental data. According to the original work [74] the behavior shown in Fig. 6d, with the ideality factor getting much greater at low temperature, should be the result of thermionic field emission (TFE). However, with a drift region doping of 4×10^{15} cm^{-3}, a calculation of the characteristic tunneling energy, E_{00} [3], suggests that no tunneling current will be present in these diodes. Another effect, known as the "T_0 anomaly", was first highlighted in this early work [75] and has been seen in other SiC diodes [23]. These effects were the first hints at inconsistency in the homogeneous TE equation and can now both be explained with an "inhomogeneous TE equation" [57].

A Richardson's plot of $log(J_S/T^2)$ versus $1000/T$ is shown in Fig. 6e. As mentioned in Section 2.2, in a homogeneous diode, a temperature independent 'effective' SBH can be determined from the slope of this Richardson's plot, while the y-intercept represents the Richardson's constant for that material. As can be seen in Fig. 6e, a linear relationship exists at the highest temperatures, where the ideality factors are very close to the unity. Here, an effective SBH of 1.17 eV can be extracted, though a Richardson constant of just 0.107 A·cm^{-2}·K^{-2} is determined, i.e., much lower that the theoretical value of 146 A·cm^{-2}·K^{-2} expected for 4H-SiC [23].

Another possible anomaly is shown in Fig. 6f, in which the temperature dependent J-V characteristics of another Mo/4H-SiC diode are shown, with non-ideal turn-on characteristics. In particular, the diode's characteristics show two turn-on regions at high temperature, and three at low temperature, suggesting multiple macroscopic regions of differing barrier height. This can happen when the Schottky contact covers a defective or dirty area of the wafer. In order to account for such a strong inhomogeneity, the prominent "non-interaction" models assume the contact to behave as multiple diodes connected in parallel, each with its own specific barrier height, effective area and series resistance.

3.2   Modelling the inhomogeneous Schottky contact

3.2.1 Non-interacting models of inhomogeneity

Over the years, several models [51,52,53,54,55,56,57] have emerged to explain the effects of a Schottky contact with barrier non-uniformity. These all involve a shift in thinking away from the notion of a single homogeneous barrier $q\Phi_{Bn}$, across the entire area A, as in the TE equation. Instead, these parallel conduction methods presume that multiple areas of different "non-interacting" barrier heights will exist at any interface, producing dominant areas of current conduction. Accordingly, the diode current will be given by the contribution from an array of i discrete regions, each with area A_i and barrier $q\Phi_i$:

$$I = A^*T^2 \exp\left[\left(\frac{qV}{kT}-1\right)\right]\sum_i A_i \exp\left(-\frac{q\Phi_i}{kT}\right) \tag{12}$$

First *Song et al.* [53] and then *Werner and Güttler* [55], assumed the existence a spatial distribution of barriers across the contact area in a real metal/semiconductor. The SBH spatial variations follow a Gaussian distribution, leading to a temperature dependence of Φ_{Bn} as follows:

$$\Phi_{Bn}(T) = \Phi_{Bn}^0 - \frac{q}{2kT}\sigma^2 \tag{13}$$

where, $q\Phi_{Bn}^0$ is the mean barrier value and σ is the dispersion of the Gaussian distribution. Both parameters can be extracted from the slope and y-axis intercept of a $\Phi_{Bn}(T)$ versus $1/T$ plot.

Table 3 reports the values of the SBH extracted by conventional $I\text{-}V$ analysis applying the TE model ($q\Phi_{Bn}$), the effective barrier extracted by $I\text{-}V\text{-}T$ analysis ($q\Phi_{Bn,eff}$), and the mean barrier ($q\Phi_{Bn}^0$) determined considering the inhomogeneity of the contact.

Table 3. Comparison of literature SBH data obtained with different analysis for various metal/4H-SiC Schottky contacts. The data are taken from Ref. [73] and references therein.

Contact	Annealing	Temperature Range (K)	Barrier from $I\text{-}V$ analysis $q\Phi_{Bn}$(eV)	Effective Barrier from $I\text{-}V\text{-}T$ analysis $q\Phi_{Bn,eff}$ (eV)	Mean Barrier $q\Phi_{Bn}^0$ (eV)
Ni	None	40 – 300	0.37–1.44	0.91	~1.65
Ni	550°C	300 – 673	~1.5	-	-
Ni	700°C	98 – 473	1.31–1.66	1.5	1.69
Au	None	50 – 300	0.93–1.18	0.98	1.24 - 1.36
Mo	500°C	298 – 498	1.01–1.07	0.9	1.14 – 1.16
W	500°C	303 – 448	1.11–1.17	1	1.28
Ti	None	173 – 373	1.24–1.27	1.22	1.31

Advancing Silicon Carbide Electronics Technology I   Materials Research Forum LLC
Materials Research Foundations **37** (2018)   doi: http://dx.doi.org/10.21741/9781945291852

Evidently, significant differences between the results obtained with the different *I-V* methods are observed. In general, the smallest barrier height values are extracted from the Richardson plots, often resulting in large discrepancies in the value of the Richardson' constant from the theoretical one. This is a direct consequence of using the nominal diode area as input data in the models.

Brezeanu et al. [73] recently proposed an analytical method, based on discrete non-interacting parallel conduction, to predict the behavior of inhomogeneous SiC Schottky diodes at high temperatures. It considers the contact as an array of *m* discrete regions, each with different barrier heights ($q\Phi_{Bn,i}$) and effective areas A_i, representing a fraction of the total contact area ($A_i = A/a_i$). Thus, the total current is written as the sum of all contributions, assuming the TE formalism [73]:

$$I = AA^*T^2 \exp\left(\frac{qV}{nkT} - 1\right) \exp\left(-\frac{q\Phi_{Bn,1}}{kT}\right) \sum_{i=1}^{m} \frac{1}{a_i} \exp\left(-\frac{q\Delta\Phi_{Bn,i}}{kT}\right) \tag{14}$$

where, $q\Phi_{Bn,l}$ is the lowest barrier present on the contact surface and $\Delta\Phi_{Bn,i} = \Phi_{Bn,i} - \Phi_{Bn,l}$ (with $\Delta\Phi_{Bn,l} = 0$).

In the proposed description, Eq. 4 is modified by introducing an effective area A_{eff} and an effective SBH ($q\Phi_{Bn,eff}$) [73]:

$$I = A_{eff} A^* T^2 \exp\left(-\frac{q\Phi_{Bn,eff}}{kT}\right) \exp\left(\frac{qV}{nkT}\right) \tag{15}$$

where $A_{eff} = A\exp(-p_{eff})$ and p_{eff} is an additional parameter introduced to quantitatively assesses the degree of non-uniformity of the barrier [73]. The value of $q\Phi_{Bn,eff}$ can be extracted from the slope of the Richardson's plot.

This characterization method was successfully tested on fabricated 4H-SiC Schottky diodes with annealed Ni and Pt contacts, exhibiting varying degrees of inhomogeneity. As an example, Fig. 7a shows the temperature dependent forward *I-V* characteristics (symbols) of a typical annealed Ni/SiC Schottky contact, with three regions, having different SBH and widely different areas (given in Fig. 7b). As can be seen, there is a good agreement between experimental data and model calculated plots (continuous lines) in the entire temperature range (-100°C – 650°C). This accurate matching was possible by dividing the entire temperature interval into three regions,(Fig. 7b). An important observation is that, at low temperatures (-100°C – 0°C), the region with the lowest barrier contributes over 99% of the total current through the contact, even if its active area is several orders of magnitude lower than the entire diode surface. The zone occupying a

Advancing Silicon Carbide Electronics Technology I
Materials Research Forum LLC
Materials Research Foundations **37** (2018)
doi: http://dx.doi.org/10.21741/9781945291852

vast majority of the contact surface will dominate the conduction only above 400°C (Fig. 7b).

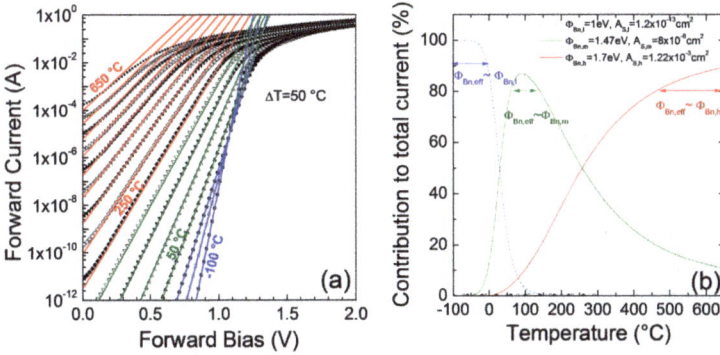

Figure 7. (a) Experimental Ni/4H-SiC Schottky diodes forward I-V characteristics (symbols) and simulated I-V curves (continuous lines) using patches with 3 different barriers applying the method proposed in Ref [73]. (b) Contribution to the total current of the different patches in the entire investigated temperature range. Reprinted from Ref. [73], with the permission of AIP Publishing.

A limitation of the Gaussian distribution methods is that the distribution of barriers is averaged over the entire contact, without considering the area of each individual patch of SBH. Furthermore, this 'non-interacting' model of inhomogeneity neglects the idea of potential pinch-off from adjacent patches. This latter concept will be discussed in detail in the next Section, devoted to the interacting models of inhomogeneity.

3.2.2 Interacting models of inhomogeneity

The work of *Tung* [54,56,57] introduced the 'interacting' model of inhomogeneity for a single phase contact. This is similar to the *Werner and Güttler* model [55] in that it considers a distribution of barriers across a single-phase contact, and sums their contribution. However, this model uses the pinch-off concept, in which the contribution of any area, or 'patch', of low SBH will be hindered (i.e., pinched-off) if it is small compared to the Debye length and it is surrounded by area of larger SBH. Pinch-off occurs when an electron passing through the space charge region, towards the interface, must pass over a potential greater than that of the patch at the surface. The electrostatics of this are visualized in Fig. 8, in which the space charge region below the interface is shown as it spans a patch of radius 30 nm with a SBH 0.4 eV less than the region that surrounds it on all sides. It can be seen that although the low barrier patch has a potential, $\Phi_B^0\text{-}\Delta$, of 0.4 eV at the surface, the 'saddle point' that forms 35 nm from the surface has a potential of 0.48 eV. The saddle point potential is dependent on patch area, doping, temperature and applied bias, giving a response that will vary greatly between diodes, and will be highly temperature and voltage dependent. The absence of pinch-off from the 'non-interacting' models of inhomogeneity, described in the previous Section, is considered as their weakness [57].

A full derivation of *Tung*'s model is beyond the scope of this book chapter, but this can be found in his papers

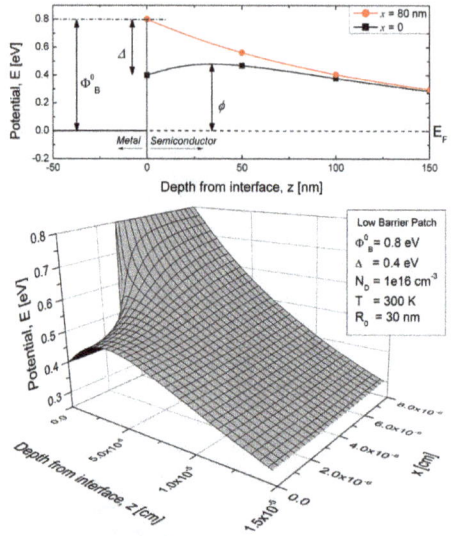

Figure 8. 2D and 3D representation of Tung's model for inhomogeneous Schottky barriers, showing how a low barrier region is pinched off by the surrounding areas of high potential. The figure is adapted from Ref. [71].

[54,56] and comprehensive reviews [57,72] of the subject. In short, however, the total current passing over the barrier is considered to be the sum of all the pinched-off (p-o) patches and all the non-pinched-off (n-p-o) patches. In full, *Tung*'s model, including the important effects of resistance is:

$$I = \sum_i I_i = A^*T^2 \sum_i^{n-p-o} A_i \exp\left(-\frac{q\Phi_i}{kT}\right)\left[\exp\left(\frac{qV}{kT} - \frac{qI_i}{kT}\frac{\rho\,t}{A_i}\right) - 1\right] +$$

$$+ A^*T^2 \sum_i^{p-o} A_{i,eff} \exp\left(-\frac{q\Phi_{i,eff}}{kT}\right)\left[\exp\left(\frac{qV}{kT} - \frac{qI_i}{kT}\frac{\rho}{4\sqrt{A_{i,eff}/\pi}}\right) - 1\right]$$

(16)

where ρ is the resistivity of the material and t is the drift region thickness.

A patch parameter γ is introduced ($\gamma = [3\Delta R_0^2/4]^{1/3}$) that relates the patch surface potential, Φ_B^0-Δ, and radius, R_0, (Fig. 8) to the effective area $A_{i,eff}$ and barrier height $q\Phi_{i,eff}$ of the patch in the space charge region. A Gaussian distribution of γ is assumed, such that

$$P(\gamma) = \frac{1}{\sigma_\gamma \sqrt{2\pi}} \exp\left(-\frac{\gamma^2}{2\sigma_\gamma^2}\right) \tag{17}$$

where σ_γ is the patch standard distribution. For each individual patch:

$$\Phi_{i,eff} = \Phi_B^0 - \gamma \left(\frac{qn_{n0}V_{bi}}{\varepsilon_0\varepsilon_S}\right)^{1/3} \tag{18}$$

$$A_{i,eff} = \frac{4\pi\gamma}{9}\frac{kT}{q}\left(\frac{\varepsilon_0\varepsilon_S}{qn_{n0}V_{bi}}\right)^{2/3} \tag{19}$$

where ε_0 and ε_s are the vacuum and the semiconductor permittivity, n_{n0} the number of carriers and V_{bi} the built-in potential.

This is the most complete model of SBH inhomogeneity to date. *Tung* has shown [56] how his detailed equations can be used to reproduce numerous effects of inhomogeneity, including *C-V* and *I-V* SBH discrepancies, high ideality factors, temperature dependent SBHs and ideality factors, voltage dependent *I-V* characteristics, and the various inconsistencies in Richardson's plots and *nkT* versus *kT* plots, including the T_0 anomaly. However, its complexity means that there are too many free parameters to attempt to use the entire equation to fit to experimental data, given that they include the resistivity (ρ), patch density, and two Gaussian distributions (each with associated average barrier heights and standard deviations) representing the pinched-off and non-pinched-off areas. This explains the continued use of the simpler *Werner and Güttler* model [55,64,73], which can be easily fit to experimental data.

Two attempts have been made to simplify the *Tung* model, and to apply it to practical SiC Schottky contacts. The first, by *Roccaforte et al.* [23] aimed to understand the effective area contributing to the conduction in an inhomogeneous SiC diode, while correcting the inherent errors within a Richardson's plot. This method simplifies Eq. 16 by considering only those patches that were pinched-off, and by assuming a single γ. This allowed $\Phi_{i,eff}$ and a number of patches N of area A_{eff} to be substituted into the conventional TE equation. Furthermore, by assuming the effective barrier as the one extracted from a Richardson's plot and Φ_B^0 as the homogeneous barrier height (i.e., extrapolated at $n = 1$ in a n versus Φ_B plot), a value for the parameter γ could be determined. This method could be fit to experimental *I-*

V results considering the total effective area of the diode NA_{eff} and to correct, accordingly, the Richardson's plot, such that a more accurate value of the Richardson's constant could be extracted (see Figs. 9a and 9b). This method suggested that the current was only passing through the areas of lowest SBH, making up around 1-2% of the original area.

A second method proposed by *Gammon et al.* [71] was used to fit Ni/SiC Schottky *I-V-T* data with a reduced version of Eq. 16. This method assumed that the majority of current conduction will come from pinched-off patches, and hence the first term in Eq. 16, the n-p-o sum, is neglected. This leaves only the second term of Eq. 16, which contains four potential fitting parameters, ρ, the density of patches (C_T), Φ_B^0 and σ_r. To reduce this, C_T was estimated to be $1 \times 10^8 \, cm^{-2}$ (a weakness of this method), and the other three parameters fitted to the data. The results, seen in Fig. 10a, show that this method could produce excellent fitting from 320 K down even to 20 K, where extremely high ideality factors have led many to speculate that the TE equation cannot still hold true.

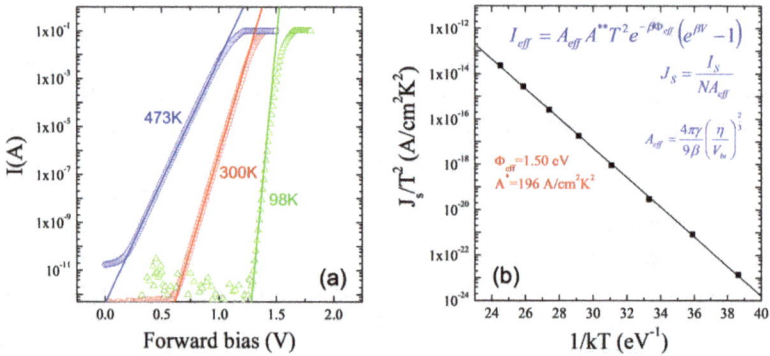

Figure 9. (a) Experimental I-V characteristics of Ni₂Si/4H-SiC Schottky diodes at three different temperatures (symbols) and fit of the data (continuous lines) using the Tung's model assuming a single γ. (b) Modified Richardson's plot of the same data from which an effective barrier height of 1.50 eV and a Richardson's constant of 196 A/(cm²K²) could be extracted. Reprinted from Ref. [23], with the permission of AIP Publishing.

Also reproduced is the curvature of the plots, the voltage dependence, across the temperature range. Finally and importantly, the barrier heights, Φ_B^0, displayed a small, almost linear, negative temperature dependence across the full temperature range as seen in Fig. 10b. This represents the temperature dependence one might expect given the bandgap and Fermi level temperature dependencies.

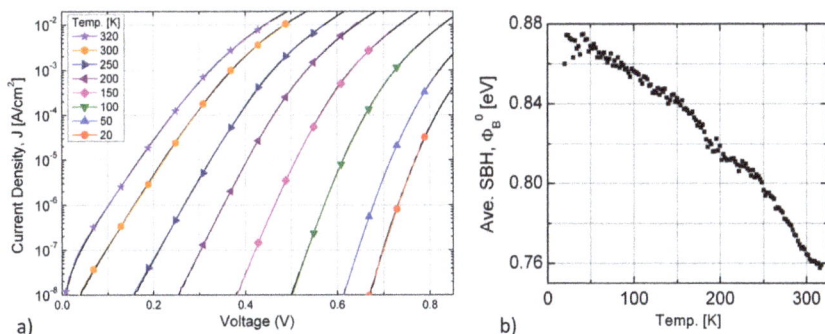

Figure 10. (a) J-V characteristics of a Ni/4H-SiC Schottky diode measured in the temperature range 20 - 320K and fitted with the reduced Eq. 16; (b) Distribution of SBHs across this temperature range, demonstrating a negative temperature dependence. The data are from Ref. [71].

3.3   Characterization of the Schottky barrier inhomogeneity at the nanoscale

To monitor the metal/SiC barrier inhomogeneity, several nanoscale resolution scanning probe microscopy methods have been proposed in the last decades.

In 1988, Bell and Kaiser demonstrated the ballistic electron emission microscopy (BEEM), a method based on scanning tunneling microscopy (STM), enabling spatially resolved carrier-transport spectroscopy at interfaces [76,77]. The energy band diagram for the BEEM setup in the case of a metal/n-type semiconductor Schottky barrier is represented in Fig. 11a. The method is based on the injection of "hot electrons" by tunneling from the negatively biased STM tip to the metal film (base) through a vacuum barrier. In the case of ultra-thin metal films, a small fraction of the hot electrons propagate ballistically in the base. When a bias greater than the metal-semiconductor SBH is applied to the tip, hot electrons are able to overcome this barrier and can be collected to the substrate terminal (collector). During BEEM measurements, the collector current (I_c) is measured at different tip positions by ramping the tip bias V_{tip}, with the tip current and base potential held constant. The I_c-V_{tip} characteristics typically exhibit a threshold voltage V_{th} for current onset, directly related to the Schottky barrier height.

This technique has been employed to map the homogeneity of Pt and Pd Schottky contacts on both 6H- and 4H-SiC (0001) [78]. The two threshold values observed in the BEEM I_C-V_{tip} characteristics of Pt and Pd contacts to 4H-SiC provided evidence of a second conduction band minimum in the 4H-SiC band structure [79].

An alternative approach for nanoscale SBH measurements based on conductive atomic force microscopy (C-AFM) has been demonstrated by *Giannazzo et al.* [80]. In this approach, given that the C-AFM operates in contact mode and the feedback for the tip height position is based on the cantilever deflection, also samples containing both conductive and insulating regions on the surface can be characterized, overcoming one limitation of BEEM. Moreover, the measured current values are typically in the order of the nA, i.e. 2-3 orders of magnitude higher than the BEEM currents. When the C-AFM tip is scanned over an ultra-thin (1-5 nm) metal film deposited on the SiC surface, the biased region of the contact will be confined within 10-20 nm (i.e., the size of the tip diameter). In this way, a "nano-Schottky diode" is formed point-by-point.

The energy band diagram for the C-AFM based setup in the case of a metal/n-type semiconductor Schottky barrier is represented in Fig. 11b. The biased conductive C-AFM tip is in contact on a nanometric area with the ultra-thin metal layer. A current sensor is

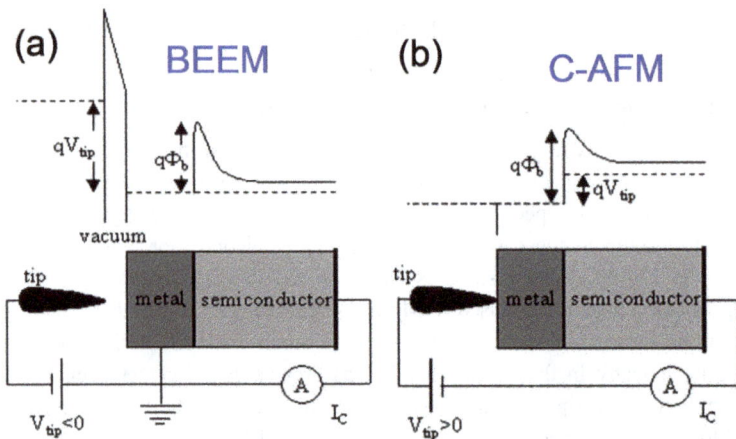

Figure 11. Energy band diagrams and schematic representations of the (a) BEEM and (b) C-AFM setups for local Schottky barrier mapping of a metal/n-type semiconductor contact.

connected in series between the tip and the back side Ohmic contact of the semiconductor, enabling the measurement of current on the nA range with a pA sensitivity.

The typical I-V_{tip} characteristics measured on a nano-Schottky diode exhibits a rectifying behavior. By fitting the forward bias I-V_{tip} characteristics measured at different tip positions with the TE model, 2D maps of the SBH can be obtained with a 10-20 nm spatial resolution and with an energy resolution in the sub-0.1 eV range.

The C-AFM approach has been applied to map the lateral distribution of the

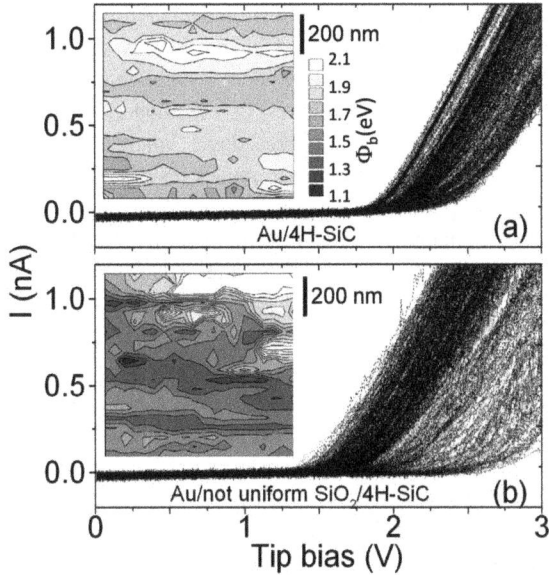

Figure 12. Monitoring the metal/SiC barrier inhomogeneity by C-AFM. Set of local I-V curves collected on arrays of 25×25 tip positions in 1 μm×1 μm area of the Au/4H-SiC (a) and Au/SiO₂/4H-SiC (b) samples. 2D maps of Φ_B evaluated by fitting of the individual I-V curves (insets of (a) and (b)). The data are taken from Ref. [80].

barriers in an Au/4H-SiC Schottky system, monitoring the effect of an ultra-thin (~2 nm) inhomogeneous SiO₂ layer between the 4H-SiC and the Au film [80]. As an example, a set of local I_c-V curves collected on an array of 25 × 25 tip positions in 1 μm × 1 μm area in the Au/4H-SiC region is shown in Fig. 12a. An identical array of curves was collected in the Au/SiO₂/4H-SiC region in Fig. 12b. Each set of 625 curves exhibits a spread, which is wider in the case of the Au/SiO₂/4H-SiC system. Two arrays of Φ_B values for the different tip positions were obtained, and 2D maps of barrier were derived from those arrays (see insets in Fig. 12). These maps clearly allow to visualize the barrier inhomogeneity in the two cases studies.

4.   High voltage SiC Schottky diodes technology

This Section describes the most relevant technological aspects related to the fabrication of Schottky Barrier Diodes (SBDs) based on 4H-SiC. In particular, a typical sequence adopted for 4H-SiC Schottky diodes fabrication is depicted, highlighting the main technical issues encountered in the various steps. Then, the concept of the Junction Barrier Schottky (JBS) is introduced, highlighting the advantages of such a device layout for the overall performances of the diode. The trade-off between the specific on-resistance (R_{ON}) and the breakdown voltage (V_B) of the devices is also presented, discussing the recent trends to reduce the R_{ON} for devices operating below 1 kV and the common approaches used to obtain efficient edge termination structures in high-voltage devices. Finally, the case of the Schottky-like SiC heterojunction diode is presented in the final part as an example of an alternative processing step to control the value of the Schottky barrier height.

4.1 Schottky Barrier Diode (SBD)

4H-SiC Schottky diodes were the first commercialised SiC devices, released in 2001. Although in principle the technology of Schottky diodes is relatively simple with respect to other devices (like MOSFETs or JFETs), there are some important differences with respect to silicon technology, which required significant technological effort to establish a reliable industrial technology flow. In fact, the fabrication of SiC diodes must take into account some requirements imposed by the physical properties of the material, e.g., extremely low diffusivity of the implanted species even at high temperatures, ion-implantation at high target temperature to minimize the formation of stable lattice defects, high-temperature post-implantation annealing for electrical activation of the dopant, high annealing temperature for Ohmic contact formation, etc..

Fig. 13 depicts a schematic cross section of a 4H-SiC Schottky Barrier Diode (SBD). As

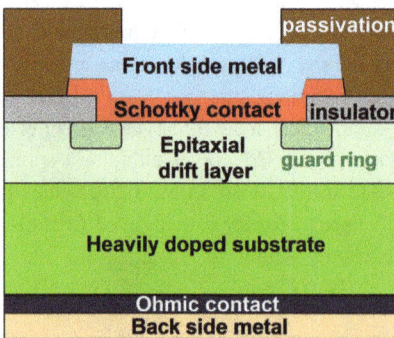

Figure 13. Schematic in cross section of a 4H-SiC Schottky Barrier Diode (SBD).

can be seen, the main building blocks of the devices are a Schottky contact on the front-side, an Ohmic contact on the back-side of the wafer, an implanted edge termination at the Schottky contact periphery and some dielectric stacks.

The sequence of the process flow to fabricate such a devices is illustrated in more detail in Fig. 14. For the fabrication of a 4H-SiC SBD, a *n*-type epitaxial layer grown onto a heavily doped 4H-SiC *n*-type substrate is required (Fig. 14a). The heavily doped *n*-type substrate has typically a concentration in the range of

$5 \times 10^{18} - 1 \times 10^{19} \text{cm}^{-3}$, corresponding to a resistivity value in the order of 20 mΩ·cm. The main specifications of the epitaxial drift layer (i.e., thickness and doping) depend on the targeted blocking voltage of the diode. As an example, for a 650 V diode, a 5 µm thick epitaxial layer with a concentration $N_D = 1 \times 10^{16} \text{cm}^{-3}$ is generally used. Such a layer is able to withstand an off-state voltage of up to 1000 V before avalanche breakdown occurs, which in turn allows a sufficient safely margin at the target voltage of 650 V.

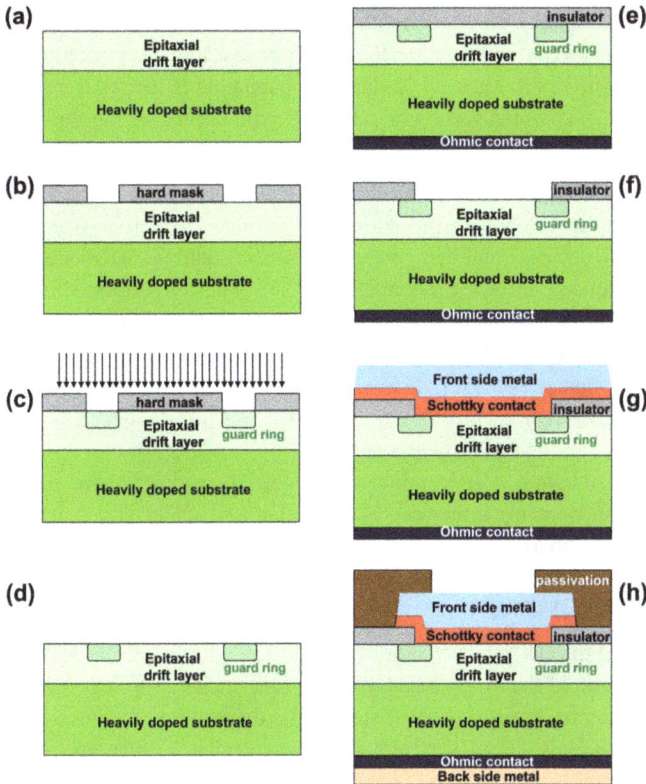

Figure 14. Schematic sequence of the process flow adopted for the fabrication of a 4H-SiC Schottky Barrier Diode (SBD).

After standard cleaning procedures, the deposition of a dielectric layer is performed on the sample front-side, and is followed by lithographic patterning (Fig. 14b), in order to define a "hard mask" for guard-ring ion-implantation. Clearly, the thickness of the dielectric hard mask has to be properly chosen to prevent implantation outside the guard-ring regions, i.e., avoiding the Schottky active area. In order to form the guard-ring area, which acts as the edge termination for the device, a low dose p-type implant is used (Fig. 14c). The implanted guard-ring prevents early breakdown by minimizing electric field crowding [81]. Ion-implantation is performed with the substrate at high temperature (400-500°C) in order to reduce the formation of ion-beam-induced defects, which would be detrimental for the electrical properties of the layer. In the early literature on high-voltage 4H-SiC Schottky diodes, the guard-ring was created by a highly resistive layer, formed either by Argon implantation [82,83,84] or by Boron implantation followed by annealing at 1050°C [85]. Later on, Al-based implantation processes were proposed [86] and are nowadays commonly used to achieve p-type doping of the guard-ring area in practical devices.

After the dielectric hard mask removal (Fig. 14d) a high-temperature thermal treatment is performed in order to activate the Al-implanted ions. Here, it is worth noting that, compared to Si devices, high-temperature treatments during the diode process flow require new solutions. In fact, these treatments are typically performed at temperatures above 1600 °C. As a consequence, significant surface degradation, also known as "step-bunching", can occur on the sample surface, especially in the regions which have undergone ion-implantation [35]. This effect can become crucial on p-type implanted regions, which may evolve into an irregular groove structure compared to non-implanted areas, exhibiting lines parallel to the steps given from the off-axis orientation of the substrate [87]. Although it is in principle possible to work on these surfaces (as the devices operation is mainly influenced by the metal/semiconductor contact fabricated on the non-implanted area), to prevent this undesired roughness, a carbon-based capping layer can be used during high-temperature annealing and then removed by specific oxidation processes [87,88,89].

After completing the electrical activation process step, a thick dielectric layer (e.g., SiO_2) is deposited on top of the 4H-SiC surface (Fig. 14e), acting as a field oxide in the edge termination structure. Then, with the wafer front-side protected by this oxide, an Ohmic contact on the heavily doped substrate is formed, typically by Nickel deposition and sintering at temperatures above 900 °C, leading to the formation of a silicide layer Ni_2Si [2,90]. The need of a high-temperature thermal budget for Ohmic contacts implies that the Schottky contact formation step is not the final one like in Si technology. In fact, in 4H-SiC, functional Schottky barriers are obtained with limited thermal budget (see Table 1) and, hence, can be formed only after the sintering of the Ohmic contacts at the back-side. This aspect will be further discussed in Section 4.3, when the need of substrate thinning technology is explained.

The following fabrication step is the active area opening (Fig. 14f), which determines the current capability of the fabricated diode. Thereafter, a metal stack (Fig. 14g) is deposited to form the anode contact and typically followed by a post-deposition annealing. As reported in Table 1, several metals have been tested as Schottky contacts to n-type 4H-SiC. However, typically low barrier metals (Ti, Mo…) are the most common ones used in the commercial devices. The metal stack is composed not only by the thin layer that forms the Schottky barrier on 4H-SiC (e.g., Ti), but also by a thicker metal overlayer (e.g., Al) that guarantees an adequate current flow through the diode and aids wire bonding.

Finally, metal definition by lithographic steps, the passivation by polyimide and back-side metallization (Fig. 14h) complete the fabrication flow, and make the diode ready for packaging and electrical characterization.

Fig. 15 shows the typical forward and reverse characteristics of a 6 A/650 V 4H-SiC SBD mounted in a standard TO220 package. As can be seen, the nominal current of 6 A is reached at a forward voltage drop V_F of approximately 1.4 V, with a specific on-resistance of the device of 1.95 mΩ·cm^2.

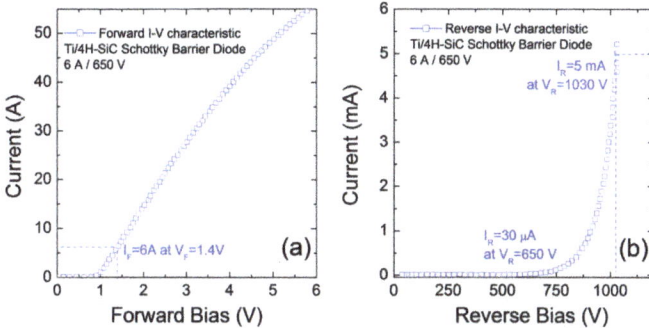

Figure 15. (a) Forward and (b) reverse I-V characteristics of a 6A/650V Ti/4H-SiC Schottky diode.

According to the TE model, the total voltage drop V_F across the Schottky diode can be written in terms of the barrier height $q\Phi_{Bn}$ and of the ideality factor n as:

$$V_F = n\frac{kT}{q}\ln\left(\frac{J_F}{A^*T^2}\right) + n\Phi_{Bn} + R_{ON}J_F \qquad (20)$$

where R_{ON} is the specific on-resistance and J_F is the forward current density.

Clearly, changing the Schottky contact (i.e., the SBH $q\Phi_{Bn}$) will directly influence V_F and, hence, the on-state power dissipation of the device. As specified in Section 2, while several metals have been employed to form Schottky contacts, Ti and Mo are currently the most popular ones in commercial 4H-SiC Schottky diodes, because of their reproducibility and low barrier height, enabling to minimize the forward voltage drop V_F and, hence, to reduce the on-state power losses of the device.

Besides the traditional metal/semiconductor approach, an alternative method to control the onset of current conduction is represented by the heterojunction diode, as will be discussed in Section 4.5.

4.2   Junction Barrier Schottky (JBS) diode

In spite of the remarkable performance obtainable with 4H-SiC SBDs, these devices exhibit some limitations both in terms of high-current capability and reverse leakage current in blocking mode. In particular, although 4H-SiC SBDs can be designed for high forward current, they present a serious limitation when operating in regime of over-load current (as occurs in some specific applications). In fact, the maximum surge forward current (I_{FSM}) of the device can lead to a significant self-heating effect of the material, thus resulting in the destruction of the device (e.g., degradation of the chip, including metallization). In addition, under reverse bias, the rapid increase of the leakage current

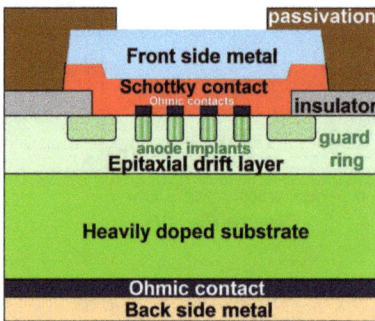

with increasing bias, due to the Schottky barrier lowering phenomenon [1,81] leads to the typical "soft-breakdown" behavior (see Fig. 15b). This increase of the leakage current is undesired especially under high temperature operation.

To overcome both limitations, the most common solution is the use of a Junction Barrier Schottky (JBS) diode. Fig. 16 shows a schematic cross section of such a device. The basic concept of the JBS is to create a potential barrier that protects the metal/SiC Schottky junction against high electric fields generated in the semiconductor.

Figure 16. Schematic in cross section of a 4H-SiC Junction Barrier Schottky diode (JBS).

This is achieved by integrating closely spaced p^+-doped regions in the Schottky contact, thus leading to the formation of parallel Schottky- and p-n junctions. Hence, in a forward biased JBS the current will flow in the un-depleted Schottky regions between the p^+-regions, preserving the unipolar operation mode. Moreover, the presence of the p^+-regions is also shown to enhance the ruggedness of the device [91]. On the other hand, under reverse bias the conduction through the Schottky regions will be suppressed by the pinch-off effect of the adjacent p-n junctions. Hence, the reverse characteristics of the JBS will resemble those of a p-n junction. As a result, the distance between the p^+-regions must be appropriately designed to optimize the trade-off between the on-state voltage drop V_F, which increases as this distance is reduced, and the leakage current, which decreases as the distance is reduced [81].

Fig. 17 shows schematically the fabrication sequence of a 4H-SiC JBS. With respect to the SBD, extra steps are required after the epitaxial layer growth (Fig. 17a). The integration of the p-n junctions in the Schottky area require the deposition and patterning of a hard mask for high-dose Al-implantation of the p^+-regions (Figs. 17b-d) and the deposition and patterning of a hard mask for lower-dose Al-implantation for the guard ring (Fig. 17e-f). After these steps, a high temperature thermal annealing for the electrical activation of these implanted regions is carried out (Fig. 17 g).

An important difference is represented by the front-side metal contact. In fact, the p^+-regions have preferably to be contacted by an Ohmic electrode, which can be different from the metal used as Schottky barrier in the un-implanted regions. As an example, nickel silicide (Ni_2Si) can be used as Ohmic contact on the p^+-region, providing reasonable values of the specific contact resistance (in the low 10^{-3} $\Omega \cdot cm^2$ range) and adequate thermal stability [92]. Clearly, the selective formation of Ohmic contacts on the p^+-regions requires at least one additional lithographic step to define the geometry (e.g., strips) followed by the same thermal annealing used to form the back-side Ohmic contact (Figs. 17h-i). After Ohmic contact formation, the fabrication of a JBS continues in a similar way as that of the SBD (Figs. 17j-l).

The main features of a 4H-SiC JBS are highlighted in Fig. 18, showing the comparison of the I-V characteristics, both in forward and reverse bias, of a JBS, a SBD and a bipolar p-n diode. In conduction mode, for low current values, the JBS diode exhibits the same onset and on-state losses of a standard SBD. On the other hand, for high current the JBS shows the same current capability of the bipolar p-n diode.

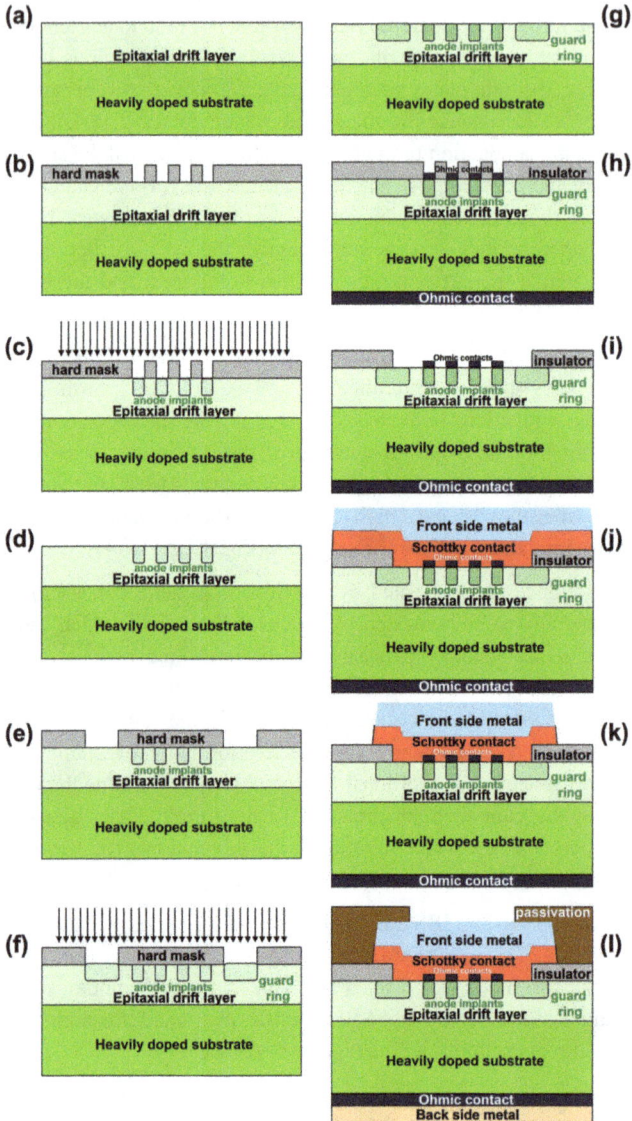

Figure 17. Schematic sequence of the process flow adopted for the fabrication of a 4H-SiC Junction Barrier Schottky diode (JBS).

Advancing Silicon Carbide Electronics Technology I
Materials Research Foundations **37** (2018)

Materials Research Forum LLC
doi: http://dx.doi.org/10.21741/9781945291852

Under reverse bias, the JBS diode characteristic is almost equivalent to that of the bipolar PN diode, showing a very low leakage current and hard avalanche breakdown almost at same voltage. This is mainly due to surface electric field reduction thanks to p-well shielding.

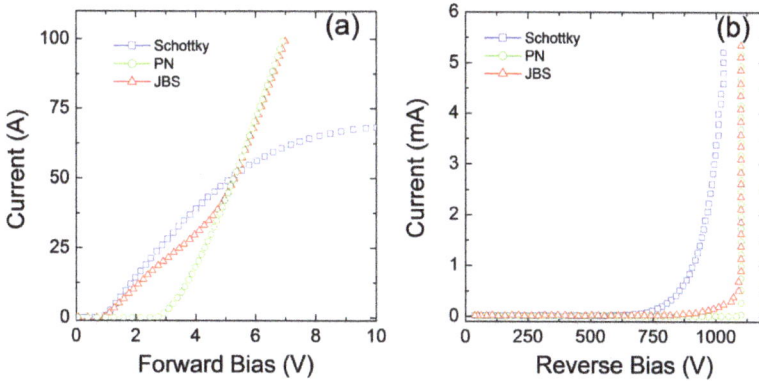

Figure 18. Comparison of the I-V characteristics in forward and reverse bias of JBS, SBD and bipolar p-n diode.

4.3 Trade-off between on-resistance (R_{ON}) and breakdown voltage (V_B)

For high-power Schottky diodes, the trade-off between the specific on-resistance (R_{ON}) and the breakdown voltage (V_B) is typically used as a measure of the device performance. In particular, to quantify the performance of power diodes in terms of the R_{ON} vs V_B trade-off, the so called *Baliga figure of merit* (BFOM) is often used, which is defined as the ratio V_B^2/R_{ON} [81].

In general, in a unipolar vertical device such as a SBD, the total R_{ON} of the diode is given by three different contributions, i.e., the resistance of the back-side Ohmic contact, the resistance of the of the drift layer and the resistance of the substrate:

$$R_{ON} = R_C + R_{drift} + R_{sub} \qquad (21)$$

The specific resistance of the Ohmic contact is typically in the order of 10^{-5}-10^{-6} $\Omega{\cdot}cm^2$ and can be neglected with respect to the other contributions to the total R_{ON}. On the other hand, as specified in the Section 4.1, the commonly used heavily doped 4H-SiC substrates have a resistivity of 20 mΩ·cm, which gives a specific resistance contribution of $7{\times}10^{-4}$ $\Omega{\cdot}cm^2$ for a substrate thickness of 350 μm. Finally, the ideal specific on-resistance of the drift layer can be expressed as [81]:

$$R_{drift} = \frac{4V_B^2}{\varepsilon_{SiC}\mu_n E_{CR}^3}$$

(22)

where V_B is the breakdown voltage, ε_{SiC} is the dielectric constant of SiC, μ_n is the electron mobility and E_{CR} is the critical electric field of the material.

Table 4 summarize the performances of a number of SiC Schottky diodes reported in the last decades in literature in terms of the R_{ON} vs V_B trade-off. The type of device is specified, including not only the standard Schottky Barrier Diodes (SBD) and Junction Barrier Schottky (JBS), but also other alternative and more complex layouts [93,100,101,105,106].

Table 4. Survey of literature data on the performances of 4H-SiC SBDs and JBS diodes.

Type of Device	R_{ON} [mΩ·cm²]	V_B [V]	BFOM V_B^2/R_{ON} [MW/cm²]	Ref.
JBS	6.9	885	113.5	[93]
PIP-JBS	7.8	1613	333.6	[93]
POP-JBS	8	1938	469.4	[93]
SBD	2.5	1200	2304	[94]
JBS	8.7	1500	1036	[95]
SBD	10.5	4600	8061	[96]
JBS	25.5	5000	3920	[97]
JBS	180	10000	2220	[98]
SBD	9	4150	7655	[99]
BC-JBS	7	1500	1284	[100]
Super-SBD	3.3	2400	6981	[101]
SBD	4.8	1300	1400	[102]
SBD	47	6000	3064	[102]
JBS	5.7	1900	2532	[102]
JBS	58	6700	3092	[102]
JBS	2.1	1400	933	[103]
JBS	7.34	3320	1501	[103]
JBS	30.3	5200	892	[103]
SBD	34	3000	2250	[104]
JBS	5.2	1252	301.4	[105]
JBS-EPL	6.3	2432	938.8	[105]
FJ-SBD	8.3	4050	1976.2	[106]

Advancing Silicon Carbide Electronics Technology I   Materials Research Forum LLC
Materials Research Foundations **37** (2018)   doi: http://dx.doi.org/10.21741/9781945291852

The literature data of Table 4 are reported in Fig. 19. The solid line represents the theoretical limit for unipolar 4H–SiC devices, determined using Eq. 22 that considers only the drift layer contribution (i.e., without the substrate). Evidently, there exists a discrepancy between the experimental data and the unipolar theoretical limit. However, while for very high voltage devices the main contribution to the R_{ON} is given by the drift layer resistance and the data points approach the theoretical limit, for blocking voltage in the range of 600-1200 V there can be always a noticeable resistivity contribution coming from the SiC substrate.

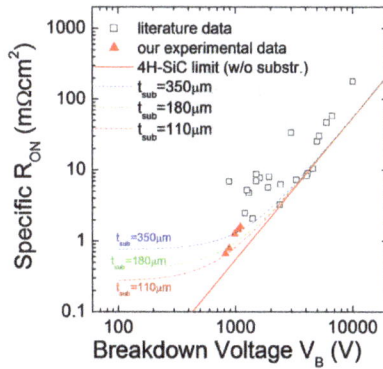

Figure 19. Trade-off plot of the specific on-resistance R_{ON} versus breakdown voltage V_B for 4H-SiC Schottky diodes. Our experimental data and some literature data from Table 4 are reported. The theoretical curves of the 4H-SiC unipolar limit without and with the substrate resistive contribution (for different substrate thickness values) are also reported.

The dashed lines reported in Fig. 19 represent the theoretical R_{ON} curves calculated including also the contributions of the substrate and of the Ohmic contact resistance for different values of the substrate thickness, i.e., 350 μm, 180 μm and 110 μm.

The single resistive contributions to the total R_{ON} in the two extreme case studies (350 μm and 110 μm) are also reported in a pie chart in Fig. 20.

For a typical substrates thickness of 350 μm and a 5 μm thick drift layer with a concentration of $N_D = 1 \times 10^{16}$ cm^{-3}, i.e., suitable epilayer specification for a 650 V diode, the resulting substrate resistivity is about 71% of the total device R_{ON}. As the heavily doped 4H-SiC substrate has practically no electrical function, in order to reduce its high resistive contribution there is the need to partially remove it by mechanical grinding, but only to a limited extent to avoid compromising the manufacturing feasibility of the wafer. As an example, in Fig. 19 and in the pie chart of Fig. 20 the advantage of reducing the substrate thickness down to 110 μm is clearly highlighted, as the substrate resistance is reduced to 44% of the total device R_{ON}.

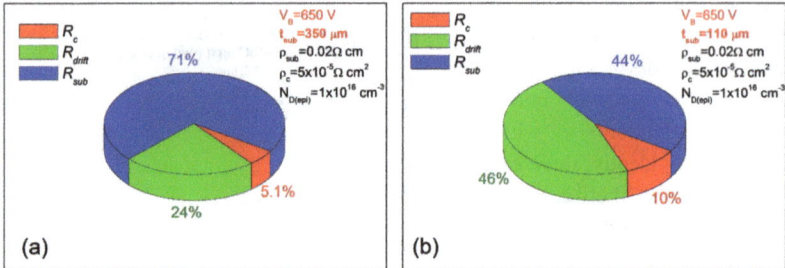

Figure 20. Pie chart of the calculated resistive contributions to the total R_{ON} of a 4H-SiC Schottky diode for two different substrate thickness of 350 μm and 110 μm. For this calculation, the following parameters have been considered: substrate resistivity of 0.02 Ω·cm, specific contact resistance of back-side contact of 5×10^{-5} Ω·cm^2, targeted breakdown voltage of 650 V, epilayer doping concentration of 1×10^{16} cm^{-3}.

Clearly, substrate thinning increases the chance of wafer breakage during device processing and, often, is preferably carried out at the end of the process flow. Such a modification of the fabrication flow requires the introduction of new solution for the backside Ohmic contact. In fact, as the backside Ohmic contact typically require large thermal budget (>900 °C), performing this process at the very end of the fabrication sequence would lead to an electrical and structural degradation of the front-side Schottky contact. For that reason, laser annealing processes suitable for nickel silicide Ohmic contact formation with negligible thermal impact on the device front-side are currently employed by some companies in the fabrication flow of 650 V SBDs [107].

4.4   Edge termination structures for 4H-SiC Schottky diodes

The theoretical reverse blocking capability of a device planar structure is primarily limited by the drift layer doping concentration level (N_D). In particular, Baliga proposed the following expression for the breakdown voltage (V_B) of 4H-SiC devices [81]:

$$V_B = 3\times10^{15} N_D^{-3/4} \tag{23}$$

However, in real devices, the blocking voltage does not reach the ideal parallel plane breakdown value, either due to the presence of "killer defects" in the device active area that cause premature failure or, more simply, because of the electric field crowding effect at the device edge [81]. In fact, in a planar structure diode, the peak of the electric field under reverse bias always occurs within the depletion region, near the electrode edges. The electric field in the depletion layer becomes spatially non-uniform and reaches its largest value where the curvature radius is the smallest. The electric field crowding can be mitigated using *edge termination* at the Schottky contact periphery. Several solutions have been reported in literature, e.g., field plate [81,108,109,110,111,112,113], highly resistive or *p*-type doped guard rings [82,83,84,85,86,], junction termination extension [114,115], etc..

Some examples of edge terminations, which can be used to optimize the breakdown of 4H-SiC Schottky diodes, are shown in Fig. 21.

The field plate is probably the simplest and easy way to fabricate an edge termination. The field plate consists in a metal overlapping on a dielectric layer, which ensures field distribution uniformity at the contact periphery (Fig. 21a). A MOS capacitor is formed at the surface of the device to spread out the field. This produces the same effect as introducing positive charge in the drift layer. Besides the reduction in electric field magnitude, the field plate also shifts the maximum of the field peak away from the contact periphery. A more efficient variant of this concept is the three step field plate termination (Fig. 21b) [4].

Brezeanu et al. [111] proposed a field plate edge termination technique based on an oxide ramp profile, with very small angle of 5° (Fig. 21c). In order to achieve the small ramp angle, oxide layers with different impurities concentrations were grown on top of the semiconductor drift layer, followed by a two steps wet etching process. Several diodes on different materials, suitable for various applications, were demonstrated employing the oxide ramp termination [111]. Common alternatives to the simple field plate are represented by implantation-based edge termination. As an example, the field plate termination can be improved by the presence of a highly resistive ring (e.g., argon or boron implanted) [82,83,84,85,110], as shown in Fig. 21d.

More complex is the multiple floating field rings, formed by concentric highly doped p^+-rings surrounding the active area of the device (Fig. 21e). In this structure, which can reach more than 90% of the ideal breakdown, the voltage is supported between each pair of guard rings. The spacing and depth of the rings are designed such as to obtain similar electric field peaks between each pair of rings. As specified in Sections 4.1 and 4.2, guard rings are widely used in 4H-SiC JBS, as they can be often formed (or annealed) simultaneously with the main junction or grids. A further option is the so-called junction termination extension (JTE), including the implantation of a *p*-type around the Schottky contact, layer in order to gradually reduce the electric field (Fig. 21f). Clearly, the optimal implantation dose of the JTE to maximize the breakdown voltage depends on the

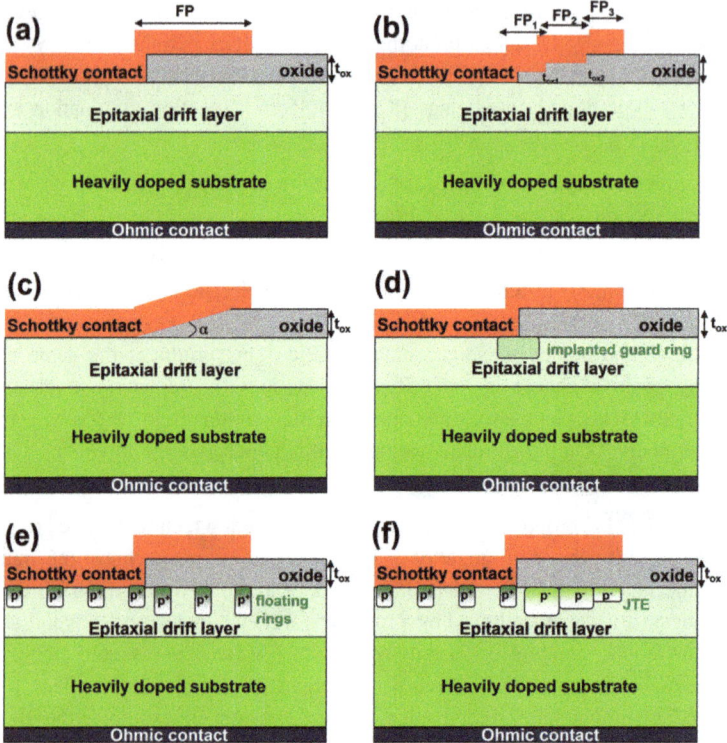

Figure 21. Examples of edge termination structures used in 4H-SiC Schottky diode technology: (a) classical field plate; (b) 3-steps field plate; (c) oxide ramp termination; (d) implanted resistive guard ring edge termination; (e) floating guard rings ; (f) multiple JTE zone.

doping concentration of the epitaxial layer. Employing a multiple JTE zones termination (Fig. 21f), with the doping level decreasing from the active area towards the edge of the device, has a similar effect to that achieved by using floating guard rings: a gradual discharge of the depletion region at the edge of the device. Besides the examples illustrated in Fig. 21, many alternative edge termination layouts have been tested on 4H-SiC Schottky diodes. As an example, mesa termination is also commonly used in combination with field plate, guard ring or JTE methods to improve the blocking capability of the diode and achieve a breakdown voltage nearly identical to the ideal parallel plane value [114].

4.5 SiC heterojunction diodes

As specified in the previous Sections, the ability to control the SBH is critical in determining the voltage drop of Schottky diodes. Besides changing the Schottky metal or the surface preparation conditions, an alternative process to control the onset of current conduction are the SiC-based heterojunctions. Several studies have been conducted into SiC heterojunctions, which can be defined as the interface between SiC and another semiconductor. A rectifying Schottky-like interface will form when a narrow bandgap semiconductor, such as silicon [116,117,118,119,120,121,122], germanium [66,123,124], 3C-SiC [125] or even graphene [126,127], is grown on, or wafer bonded to, a low doped epitaxial layer of 4H-SiC. In theory, the benefit of a heterojunction diode over a regular metal/semiconductor contact is that the doping of the narrow bandgap layer can be used to control the barrier height at the interface. Fig. 22a shows the band diagram of two extreme, ideal cases that consider only the work function of the materials, ignoring surface charge: the n^+/n^- and p^+/n^- Si/SiC heterojunction before and after contact is made. Both can be seen to make a one-sided Schottky-like rectifier, the barrier height of which is greater in the p^+/n^- diode than in the n^+/n^- diode. Despite the presence of the p^+-Si layer, both diodes can be characterized as n-type Schottky diodes as the band alignment in the SiC results in a valence band offset that is too high for hole injection.

In reality, the case of Si/SiC interface is particularly poor, as a lattice mismatch of 19.8% (SiC [0001] to Si [111]) leaves little prospect of forming a uniform, single crystal contact layer to the SiC surface. In fact, Si layers grown by MBE at 900 °C onto 4°-off-axis 4H-SiC substrates exhibit a rough morphology (see Fig. 22b), characterized by numerous merged islands of crystalline Si with different orientation [116,117,118]. Alternatively, using low temperature growth, a uniform coverage and a better morphology can be achieved. This can be seen in the AFM image of Fig. 22c for a Ge/SiC interface grown at 200˚C, where 300 nm of Ge is uniform enough that the lines visible are the step bunching lines from the SiC beneath. At raised temperature, the coverage of a Ge layer on SiC is somewhat better than for Si [66,123,124], despite an even larger 23.1% lattice mismatch. Fig. 22d, shows that it is possible to produce a relatively smooth layer of polycrystalline Ge on SiC at 500 °C. This is brought about, it is suggested, by the partial off-axis alignment of the [111] Ge surface and the [0001] SiC surface [124].

A possible alternative to MBE or CVD growth is to wafer bond the two semiconductors [119,120,121,122]. This results in an ideal crystalline top layer, but often at the expense of interface quality, which can contain an interfacial oxide, voids or processing residue. Therefore, diodes formed with conduction across this interface often displays poor turn

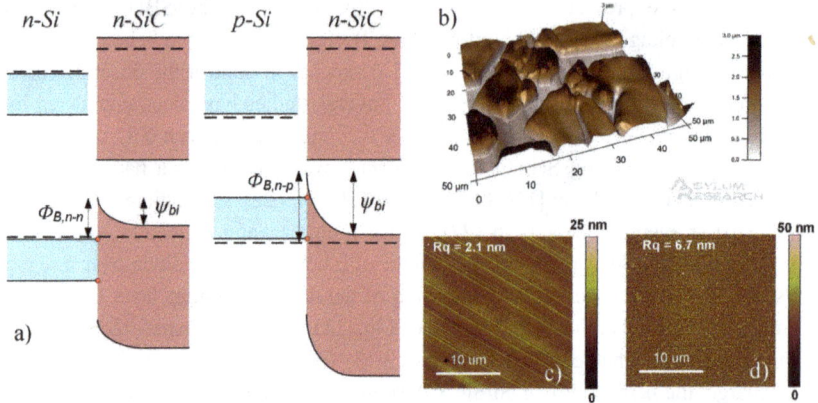

Figure 22. (a) The ideal band alignment for Si/SiC heterojunction diodes, suggesting an element of SBH control.(b) A 3D AFM image of a Si layer grown on SiC at 900 °C. c) An AFM image of the surface of a 300 nm Ge layer grown on SiC at 200 °C and in d) at 500 °C.

on characteristics including high ideality factors and resistance [120]. However, surface activated bonding, plus rapid thermal annealing at 1000 °C has been shown [122] to improve the interface, resulting in low resistance contacts with minimal reverse leakage.

Whether wafer bonded or grown, interface quality, polycrystalline layers and grain boundaries can all have a detrimental effect on the ideal band alignment presented in Fig. 22a. These effects all increase the amount of charge residing at the heterojunction interface, leading to Fermi level pinning [124]. Hence, the influence of the semiconductor doping on the Schottky barrier height having been weakened. This is demonstrated in Fig. 23, in which p^+ and n^+ Ge layers are formed on SiC at high and low deposition temperatures [124]. A good quality Ge/SiC interface exists, with the diodes all having low leakage and an ideality factor, n, less than 1.1. However, they also show the effects of Fermi pinning, given that there is only approximately 30 meV between barrier energies of the n^+ and p^+ MBE grown layers. The evident difference in resistance between the layers is due to the resistance of the Ge layer being an order of magnitude lower in the p^+

sample due to the specific dopant species used. Fermi level pinning has also been seen in Si wafer bonded heterojunction diodes [122], where the SBH remains 0.92 eV regardless of doping, and the resistance of the p^+ contact is lower than in the n^+ layer.

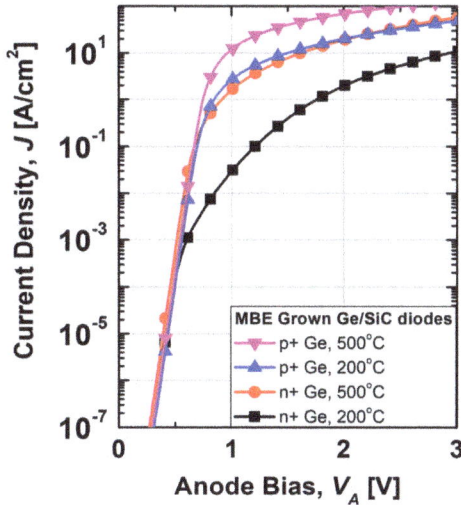

Figure 23. The electrical characteristics of MBE deposited Ge/SiC heterojunction diodes. Reprinted from Ref. [124], with the permission of AIP Publishing.

Despite the challenge in perfecting the promise of SiC heterojunction diodes, one notable example of a successfully implemented heterojunction diode is that from *Tanaka et al.* [128]. Here a low pressure CVD grown p^+-Si layer was shown to block the same voltage as a conventional SBD but with lower reverse leakage, and lower turn on voltage.

5. Examples of SiC Schottky diodes applications

The last Section of this chapter illustrates some common examples of 4H-SiC Schottky diodes applications. In particular, it will be shown that, nowadays, the most important market-driven applications are related to energy efficient power conversion systems. However, other applications (like temperature sensors, UV-detectors, etc.) are rapidly emerging not only for niche markets but also as integrated accessories in consumers electronics products. In addition, there are many other possible uses of 4H-SiC SBDs, such as radiation detectors for nuclear and space application, gas sensors, microwave circuits, etc., which will be not mentioned in this Section. More details on these specific cases can be found elsewhere [34,129,130].

5.1 Applications in power electronics

Power conversion systems are present in everyone's day by day life. Hence, the reduction of the global energy consumption is strictly related to the development of new energy efficient power devices [131]. In this context, SiC is today the most promising material to satisfy such challenging request.

The diode is widely used as a companion of the transistor in almost all the conversion systems and the addressed market is huge. As can be seen in the power versus voltage chart depicted in Fig. 24, the most common applications in the present market (e.g., in consumer electronics, renewable energies, industrial and automotive sectors, etc.) require devices able to sustain off-state voltages in the range 650 V – 1.7 kV.

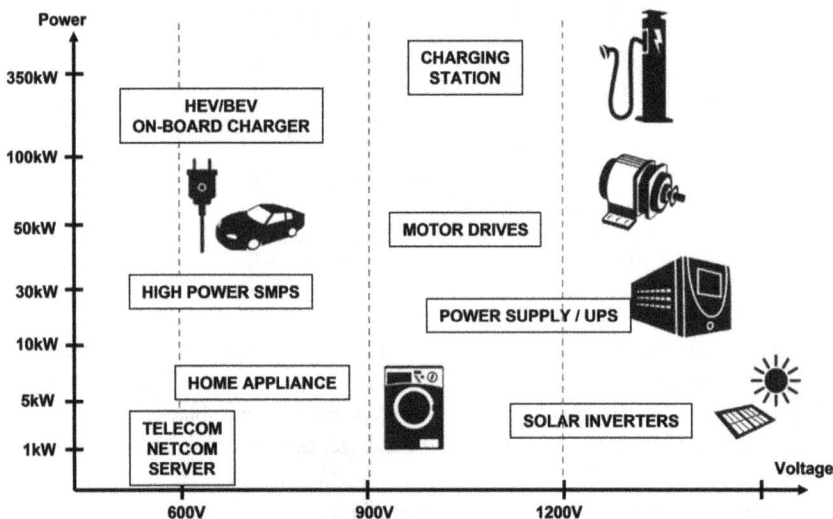

Figure 24. Power versus blocking voltage chart of the most common applications of power devices in the range 650-1700 V.

For all of these applications, the possible solution based on Silicon rectifiers is the bipolar diode, which is characterized by very high switching losses. In Fig. 25 a typical reverse

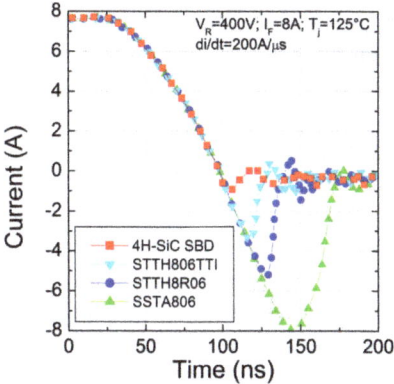

Figure 25. Reverse recovery waveforms of a 4H-SiC Schottky diode (600 V/8 A) at 125 °C compared with different ultra-fast commercial Si bipolar diodes.

recovery waveform of a 4H-SiC SBD is compared with different commercial Si bipolar diodes, each with a recovery time determined by a different carrier lifetime enhancement technique. Independent of the bipolar Si technology, the unipolar 4H-SiC SBD has the minimum recovery loss due to its absence of minority carriers. It is the fundamental characteristics of SiC, its high critical electric field, that permits a SiC unipolar diode to be rated at the same voltage as the Si bipolar device, without suffering great conduction losses.

A typical example of 4H-SiC Schottky rectifier application is represented by the Power Factor Corrector (PFC) circuit. All electronic devices require power supplies to convert the AC voltage from the grid to DC voltage for electronics (e.g., computers, telecom, etc.). Linear power supplies, even the ones with passive filtering, have a power factor (i.e., the ratio of the real power in the load to the apparent power in the circuit) lower than 1 that introduces harmonic currents into the system. The PFC boost converter is a circuit that can be added to power supplies to improve their power quality, enabling the AC input line to see near-unity power factor.

A simplified schematic of a PFC boost converter circuit is depicted in Fig. 26, comprising a diode (D) and a MOSFET (M), besides the other passive elements. During the switching operation of a PFC circuit, when the diode is turning OFF and the MOSFET is turning ON, the reverse recovery current from the diode will flow into the MOSFET, in addition to the rectified input current. Hence, to

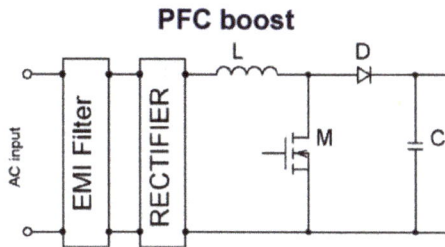

Figure 26. Simplified schematic of a Power Factor Correction (PFC) circuit.

sustain this large inrush current, a large MOSFET is required. Moreover, these switching losses also limit the frequency of operation, and the efficiency of the circuit. Therefore, a diode that would permit a higher operating frequency would allow a reduction in the size of the passive components, while near-zero reverse recovery would guarantee a higher efficiency of the PFC circuit, helping to comply with the current legal requirements. For this purpose, the near-zero reverse recovery 4H-SiC SBD (see Fig. 25) provides low switching losses, while still exhibiting comparable on-state performance as those of conventional Si rectifiers. The replacement of the conventional Si bipolar diode with a 4H-SiC SBD, even maintaining the same Si MOSFET, allows the increase of switching frequency from 10 kHz to as much as 100 kHz, resulting in the subsequent reduction in size of the inductor. In addition, the lower recovery charge allows a reduction in size of the electro-magnetic noise filter and a physically smaller MOSFET and diode. Finally, the higher temperature capability of SiC allows a reduction in heat-sink size. Clearly, the overall impact is a simplification and cost reduction of the system (even if the cost of the discrete SiC diode is higher that Si diode), leading to a much higher system conversion efficiency.

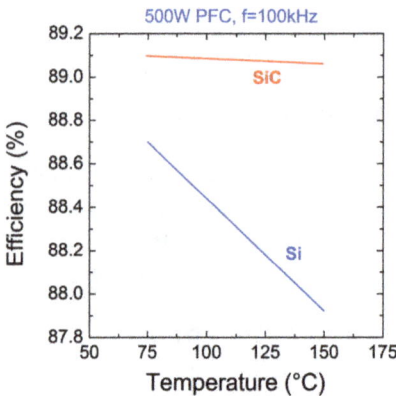

Figure 27. Comparison of the efficiency, as a function of temperature, of a 500 W/100 kHz PFC circuit based on Si devices with that of the same PFC obtained replacing the Si bipolar diode with a SiC Schottky diode.

These benefits are clearly shown in Fig. 27, which reports the efficiency, as a function of temperature, of a 500 W PFC operating at 100 kHz, where the Si bipolar diode has been simply replaced by a 4H-SiC Schottky diode, while keeping the same Si MOSFET. The efficiency is higher at an operating temperature of 75 °C with the benefit of SiC increasing with operating temperature, a result of the huge increase of losses in bipolar Si rectifiers.

To compare the performance of different 4H-SiC Schottky diode technologies in power electronics applications, both efficiency and robustness have to be taken into account. The efficiency is generally linked to the forward voltage drop at the nominal current, which is strictly related to the diode conduction losses. The robustness is given by the diode capability to withstand high current in surge mode (I_{FSM}). Fig. 28 compares the performances of latest generation 1.2 kV 4H-SiC Schottky diode technologies available on the market, reporting the ratio between the I_{FSM} and the nominal current I_F (at room temperature) as a function of the forward voltage drop V_F. In this graphical representation, the efficiency of the diode increases with decreasing V_F, while its robustness is improved with increasing I_{FSM}/I_F ratio. Clearly, in the choice of the diode design, there exists a trade-off between the low V_F value and the high I_{FSM} performances. In order to consider this trade-off, specific technological solutions (wafer thinning, improved packages in thermal performance etc.) have to be taken into account to have at the same time high efficiency and high robustness. In the near future, efficiency and robustness in the same diode technology will allow a tremendous improvement in PFC performance, producing larger opportunities in the market. In particular, a further reduction of diode area will be possible, reducing costs and allowing the use of SiC diode in high-volume low-cost applications.

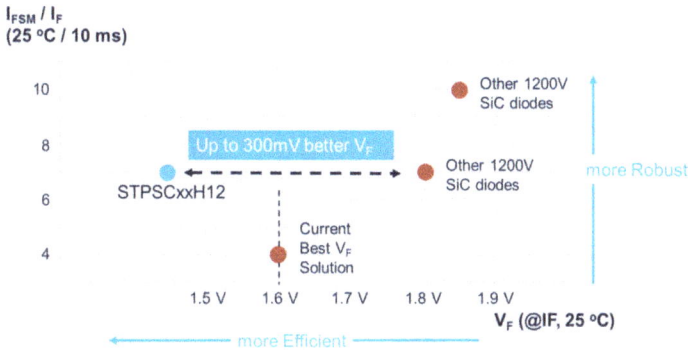

Figure 28. Graphical representation of the trade-off between efficiency and robustness for commercial 4H-SiC Schottky diodes, in a plot of the ratio I_{FSM}/I_F as a function of the V_F at room temperature.

5.2   Temperature sensors

4H-SiC Schottky barrier diodes can be used also as a high-temperature sensor, suitable in harsh environments - high shock or intense vibration, high radiation, erosive and corrosive conditions. In particular, temperature probes based on 4H-SiC SBD, capable of operating under those extreme conditions, can have significant applications in several fields, e.g., automotive and aircraft engines, geothermal systems, industrial furnaces, oil and gas detection, etc. [73,132,133,134,135,136].

For temperature monitoring sensor applications, the SiC SBDs are forward biased as a constant current. The I_F-V_F characteristics of the devices exhibit an excellent linearity of the $ln(I_F)$ versus V_F plot over many orders of magnitude, up to high temperatures. Therefore, for low current density where R_{ON} can be neglected, a quasi-linear forward voltage versus temperature dependence can be written as:

$$V_F(T) = n\Phi_{Bn} - \left[n\Phi_{Bn} + 2n\frac{kT_0}{q}\ln\left(\frac{T}{T_0}\right) - V_F(T_0) \right]\frac{T}{T_0} \tag{24}$$

where T_0 is a reference temperature and $q\Phi_{Bn}$ and n are the Schottky barrier and the ideality factor. The above equation was derived from the TE expression, written at two different temperatures (i.e., T and a reference T_0).

The detection sensitivity of the sensor, S, is defined by:

$$S = \frac{dV_F(T)}{dT} = -\left\{ n\Phi_{Bn} + 2n\frac{kT_0}{q}\left[\ln\left(\frac{T}{T_0}\right) + 1 \right] - V_F(T_0) \right\}\frac{1}{T_0} \tag{25}$$

Clearly, both $q\Phi_{Bn}$ and n should ideally be temperature-independent, in order to ensure a stable and reproducible detection sensitivity. Moreover, high SBH metals are preferred to operate at high temperatures, e.g., Ni_2Si, Pt, etc. (see Table 1).

A typical V_F-T dependence for a Ni_2Si/4H-SiC SBD temperature sensor is plotted in Fig. 29a, at several forward current levels. For a given I_F value the sensor showed very good linearity over the whole temperature range (27-400 °C). The detector sensitivity S, extracted from the $V_F(T)$ plots of two different devices (i.e., #7 and #3), is reported in Fig. 29b. As can be seen, S ranges from between 1.5 to 2.9 mV/K for a wide current variation (10 nA – 1 mA). The calculated curves were obtained taking into account the temperature dependence of $q\Phi_{Bn}$ related to the barrier inhomogeneity [73].

Typically, a temperature probe comprises a sensor diode, packaging and a processing circuit to transfer information from the sensor to the outside world. For instance, packaging solutions based on the wire bond and pressure contact technologies can be used for SiC SBDs-based high temperature sensors [132].

A standard processing circuit of the sensor's output signal (Fig. 30a) consists of an excitation circuit (the constant current source, $I_1=100 \mu A$), an offsetting block (another current source, $I_2=100 \mu A$) and an amplifier (dual single-supply operational amplifiers U_1, U_2). The voltage to current (4-20 mA) converter (U_3) is required for connection with the standard industrial acquisition system [137]. The temperature probe performance is indicated by an excellent linearity of the output current as a function of temperature, as shown in Fig. 30b.

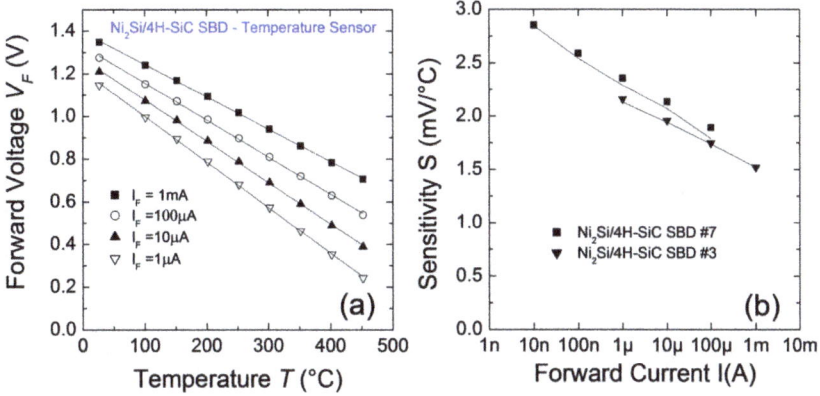

Figure 29. Application of Ni/4H-SiC SBD as temperature sensor: (a) Sensor output signal (V_F) versus temperature T for different values of forward current; (b) Detector sensitivity S for two different devices. The continuous lines in Fig. 29a are the fits of the data with Eq. 24. Figures adapted from [132] and reprinted with permission of Trans Tech Publications.

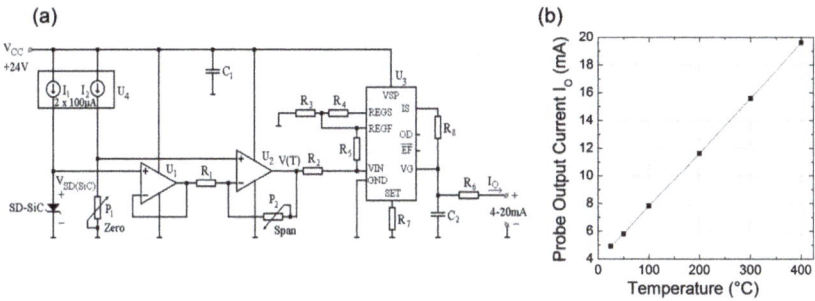

Figure 30. (a) Schematic of the processing circuit of a temperature probe based on a 4H-SiC SBD (adapted from [137]); (b) Output current of a high- temperature probe with a 4H-SiC SBD sensor.

The advantage of SBD temperature sensors, if compared with other sensors that can be on-chip integrated, e.g., thermistors, are the compatibility with IC technology, the low manufacturing costs, the quasi-linear output characteristic, preserving at the same time a high sensitivity [135]. To date, the diodes are realized for applications at temperatures of about 300 °C, i.e., oil and gas exploration, nuclear environments and similar [134].

5.3   UV-detectors

An important application field of SiC Schottky diodes is the ultra-violet (UV) radiation detection. The operating principle of this device is the detection of the photocurrent generated in the depletion region of a reverse biased SBD under UV-light exposure.

The ultraviolet (UV) region covers the wavelength range 100–400 nm and is divided into three bands: ultraviolet light A (UV-A) (315–400 nm), ultraviolet B-band (UV-B) (280–315 nm), and ultraviolet light C (UV-C) (100–280 nm). Today, UV-radiation detection has become very important for human healthcare, due to an increase in the occurrence of skin diseases all over the world [138].

Traditionally, Silicon photomultipliers have been used for the detection of UV light [139]. However, the main problem of using Si for UV-radiation detection lies in its narrow band gap (1.12 eV), leading to the need for supplementary filters to eliminate the visible and infrared components of the light, which do not need to be detected. In addition, their low quantum efficiency in the UV range, large size, high cost, and high operation voltage limit their practical use in several cases.

4H-SiC due to its large band gap (3.2 eV) means that this material will only respond to the radiation with wavelength below approximately 400 nm. The longer wavelengths from the visible and infra-red spectrum cannot be absorbed and, hence, the detectors based on SiC are insensitive to this portion of the spectrum. This characteristic is extremely advantageous since it allows SiC detectors to be used even in the presence of visible and infrared background, as occurs in many applications. In addition, owing to the low intrinsic carrier concentration of the material (in the order of $10^{-7}\,cm^{-3}$ at room temperature [140]), 4H-SiC Schottky diodes have an extremely low leakage current, thus increasing the sensitivity of the devices.

One of the first successful approaches using 4H-SiC Schottky diodes as a UV-detector was presented by *Yan et al.* [141], who fabricated 4H-SiC Schottky diodes employing an ultra-thin (7.5 nm) semitransparent Pt Schottky contact as the anode electrode. Owing to its high work function, the Pt contact provided a high SBH ($q\Phi_B$=1.52 eV) and extremely low leakage current (1.2×10^{-14} A at a reverse bias of -1 V). These devices exhibited a maximum external quantum efficiency (QE) of about 37% between 240 and 300 nm.

However, a 7.5 nm thick Pt electrode absorbs about 50% of the incident photons at 300 nm. In addition, such thin metal layers may not provide a good "intimate" contact with SiC, thus leading to reliability problems under long-term operation and/or in harsh environmental conditions. For that reason, alternatives to the ultra-thin semitransparent metals have been investigated.

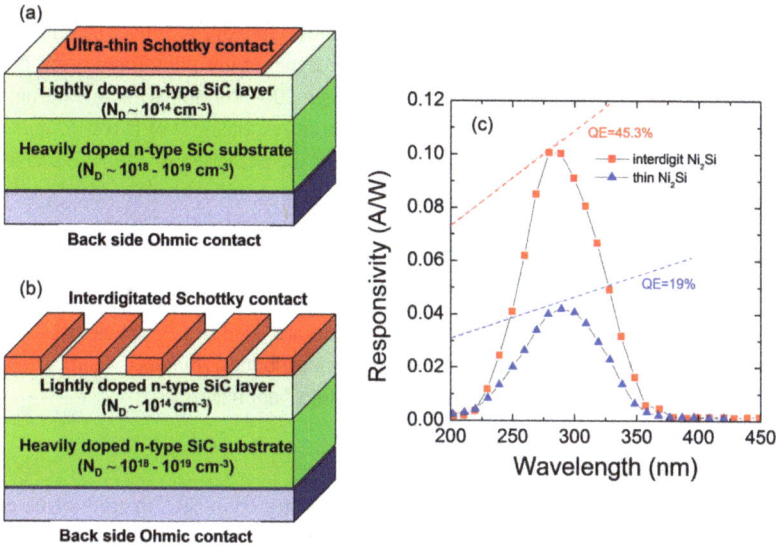

Figure 31. Schematics of a 4H-SiC Schottky UV-detectors with semitransparent (a) ultra-thin and (b) interdigitated Schottky contacts. (c) Examples of responsivity curves as a function of the wavelengths of UV-detectors fabricated using interdigitated and thin Ni_2Si Schottky metal as semitransparent electrode. The data are taken from Refs. [143,144,145].

As already reported in Section 2, nickel silicide (Ni_2Si) formed by the thermal reaction of nickel films on SiC can result in a robust metallization with a high SBH value in the order of 1.60 eV [23]. Hence, although this contact is less suitable for power devices applications, it can be promising for other applications, including the UV sensor. In particular, the low leakage guaranteed by the high SBH value and the possibility to have a "self-aligned" process (i.e., the local formation of Ni_2Si defined by a selective etch of unreacted Ni regions), can be useful for the realization of semi-transparent "interdigitated" Schottky electrodes on 4H-SiC for UV-detection. Using this concept, *Sciuto et al.* [30,142] proposed a semitransparent interdigitated Ni_2Si Schottky metal for

4H-SiC Schottky UV-detectors based on the pinch-off surface effect. The interdigitated contacts allow the active region of the device to be directly exposed to the UV radiation, thus potentially enabling a high quantum efficiency with respect to a continuous metal. Figs. 31a-c show schematically two different types of Ni_2Si/4H-SiC Schottky UV-detector as well as the responsivity curves (the generated photocurrent per incident optical power) acquired on these devices.

In particular, by appropriately selecting the thickness and doping level of the n-type epilayer, use of the interdigitated metal resulted in a maximum photon detection at 0 V (i.e., in photovoltaic regime), in the wavelength range 200–380 nm, with an external QE of about 45% at 290 nm (corresponding to a measured responsivity of 0.106 A/W at this wavelength) [143,144]. Typically, in order to obtain the depletion of the active layer in the absence of bias (photovoltaic regime), very low doping epilayers ($N_D \sim 10^{14} \, cm^{-3}$) are used. On the other hand, using a continuous Ni_2Si metal layer (about 20 nm thick) as the Schottky electrode leads to a simplification of the fabrication process [145]. However, in this case, the maximum external QE in photovoltaic regime at 290 nm was 19%.

These detectors are now commercially available and are used in several fields, e.g., for UV Index measurements in wearable devices, smartphones, tablets, weather station equipment and many other emerging "internet of things" applications.

6.   Conclusion

The Schottky diode was the first power device demonstrated in SiC. In spite of its early commercialization in 2001, the physics and technology of these devices has been investigated continuously by the academic and industrial SiC communities.

This book chapter reviewed the physics and technology of Schottky contacts to SiC. In particular, several metals have been studied over the years as Schottky barriers to n-type SiC epitaxial layers. Among them, Titanium remains the most widely used metal for high-voltage Schottky diodes, due to its low barrier height ($\sim$1.2 eV), ensuring a moderate forward voltage drop and, hence, a low power dissipation. Nickel and nickel silicide (N_2Si) are also often adopted as Schottky barrier metals for 4H-SiC Schottky diodes, although the higher barrier height values ($\sim$1.6 eV) make these systems more suitable for other applications (e.g., temperature sensors, UV-detectors, etc.).

While the thermionic emission model is used to describe the forward I-V characteristics of SiC Schottky diodes, several anomalies, often observed, in the characteristics of the diodes can only be explained by considering Schottky barrier inhomogeneity, both at a macroscopic level and at the nanoscale. Many models have been proposed to explain these effects, all involving a shift in thinking away from the notion of a single homogeneous barrier across the entire diode area. Instead, these models typically presume that multiple areas of different size and barrier heights exist at the metal/SiC interface. In this context, only advanced characterization methods can accurately monitor and quantify the degree of homogeneity of a Schottky barrier and fully predict the device behavior.

Today, reliable SiC diodes are commercially available in a wide range of voltage and current ratings, and they have been applied in power supplies and energy conversion systems, demonstrating significantly higher efficiencies compared to Si diodes. Although the technology of the SiC Schottky diode is relatively simple with respect to other devices, its consolidation required remarkable technological efforts to establish a reliable industrial technology flow. In fact, the fabrication of SiC diodes had to take into account several requirements imposed by the physical properties of the material (e.g., low diffusivity of the implanted species even at high temperatures, ion-implantation at high temperature, high-temperature post-implantation annealing for electrical activation of the dopant, high annealing temperature for Ohmic contact formation, etc.).

Now, after more than two decades of fundamental research and technological development, SiC Schottky diodes have entered our daily life, not only owing to their applications in power electronics systems, but also for their common use in wearable devices, smartphones, tablets, weather station equipment and many other emerging "internet of things" applications. However, studying the Schottky contacts remains an intriguing research topic for understanding the carrier transport mechanism at metal/SiC interfaces and, ultimately, to optimize the performance of the devices towards a further increase in their energy efficiency.

References

[1]   E.H. Rhoderick, R.H. Williams, *Metal-Semiconductor contacts*, Oxford Science Publications, Oxford, 1988.

[2]   F. Roccaforte, F. La Via, V. Raineri, Ohmic Contacts to SiC, in "SiC Materials and Devices", M.S. Shur, M. Levinshtein, S. Rumyantsev edt., pagg. 77-116, World Scientific, Singapore (2006). ISBN 981-256-835-2.

[3]   D.K. Schroder, Semiconductor Material and Device Characterization, John Wiley & Sons, Inc., Third Edition, Hoboken, New Jersey, 2006.

[4]   M. Satoh, H. Matsuo, Mater. Sci. Forum 527-529 (2006) 923-926.
       https://doi.org/10.4028/www.scientific.net/MSF.527-529.923

[5] G. Constantinidis, J. Kuzmic, K. Michelakis, K. Tsagaraki, Solid-State Electronics 42 (1998) 253-256. https://doi.org/10.1016/S0038-1101(97)00224-4

[6] P. Shenoy, A. Moki, B.J. Baliga, D. Alok, K. Wongchotigul, M. Spencer, Technical Digest., International Electron Devices Meeting (IEDM '94), San Francisco, USA, 11-14 December 1994, pp. 411-414. https://doi.org/10.1109/IEDM.1994.383380

[7] J. Eriksson, M.H. Weng, F. Roccaforte, F. Giannazzo, S. Leone, V. Raineri, Appl. Phys. Lett. 95 (2009) 081907. https://doi.org/10.1063/1.3211965

[8] J.R. Waldrop, R.W. Grant, Appl. Phys. Lett. 62 (1993) 2685-2687. https://doi.org/10.1063/1.109257

[9] J.R. Waldrop, R.W. Grant, Y.C. Wang, R.F. Davis, J. Appl. Phys. 72 (1992) 4757-4760. https://doi.org/10.1063/1.352086

[10] S.K. Lee, C.M. Zetterling, M. Östling, I. Åberg, M.H. Magnusson, K. Deppert, L.E. Wernersson, L. Samuelson, A. Litwin, Solid-State Electronics 46 (2002) 1433-1440. https://doi.org/10.1016/S0038-1101(02)00122-3

[11] F. Roccaforte, F. La Via, A. La Magna, S. Di Franco, V. Raineri, IEEE Transactions on Electron Devices 50 (2003) 1741-1747. https://doi.org/10.1109/TED.2003.815127

[12] M.O. Aboelfotoh, C. Fröjdh, C.S. Petersson, Phys. Rev. B 67 (2003) 075312. https://doi.org/10.1103/PhysRevB.67.075312

[13] F. Roccaforte, F. La Via, V. Raineri, F. Mangano, L. Calcagno, Appl. Phys. Lett. 83 (2003) 4181-4183. https://doi.org/10.1063/1.1628390

[14] H.J. Im, Y. Ding, J.P. Pelz, W.J. Choyke, Phys. Rev. B 64 (2001) 075310. https://doi.org/10.1103/PhysRevB.64.075310

[15] M. Bhatnagar, P.K. McLarty, B.J. Baliga, IEEE Electron Device Lett. 13 (1992) 501-503. https://doi.org/10.1109/55.192814

[16] D. Perrone, M. Naretto, S. Ferrero, L. Scaltrito, C.F. Pirri, Mater. Sci. Forum 615-617 (2009) 647-650. https://doi.org/10.4028/www.scientific.net/MSF.615-617.647

[17] T. Nakamura, T. Miyanagi, I. Kamata, T. Jikimoto, H. Tsuchida IEEE Electron Device Lett. 26 (2005) 99-101. https://doi.org/10.1109/LED.2004.841473

[18] K.J. Choi, S.Y. Han, J.L. Lee, J. Appl. Phys. 94 (2003) 1765-1768. https://doi.org/10.1063/1.1581347

[19]   K.V. Vassilevski, A.B. Horsfall, C.M. Johnson, N. Wright, A.G. O'Neill, IEEE Transaction on Electron Devices 49 (2002) 947-949. https://doi.org/10.1109/16.998610

[20]   S.K. Lee, C.M. Zetterling, M. Östling, J. Appl. Phys. 87 (2000) 8039-8044. https://doi.org/10.1063/1.373494

[21]   S. Nigam, et al., Appl. Phys. Lett. 81 (2002) 2385-2387. https://doi.org/10.1063/1.1509468

[22]   D.T. Morisette, J.A. Cooper, M.R. Melloch, G.M. Dolny, P.M. Shenoy, M. Zafrani, J. Gladish, IEEE Transactions on Electron Devices 48 (2001) 349-352. https://doi.org/10.1109/16.902738

[23]   F. Roccaforte, F. La Via, V. Raineri, R. Pierobon, E. Zanoni, J. Appl. Phys. 93 (2003) 9137-9144. https://doi.org/10.1063/1.1573750

[24]   A. Itoh, T. Kimoto, H. Matsunami, IEEE Electron Device Lett. 16 (1995) 280-282. https://doi.org/10.1109/55.790735

[25]   V. Saxena, J.N. Su, A.J. Steckl, IEEE Transactions on Electron Devices 46 (1999) 456-464. https://doi.org/10.1109/16.748862

[26]   D.H. Kim, J.H. Lee, J.H. Moon, M.S. Oh, H.K. Song, J.H. Yim, J.B. Lee, H.J. Kim, Solid-State Phenomena 124-126 (2007) 105-108. https://doi.org/10.4028/www.scientific.net/SSP.124-126.105

[27]   L. Huang, D. Wang, Jpn. J. Appl. Phys. 54 (2015) 114101. https://doi.org/10.7567/JJAP.54.114101

[28]   C.K. Ramesha, V. Rajagopal Reddy, Superlattices and Microstructures 76 (2014) 55-65. https://doi.org/10.1016/j.spmi.2014.09.026

[29]   F. Roccaforte, F. La Via, V. Raineri, P. Musumeci, L. Calcagno, G.G. Condorelli, Appl. Phys. A 77 (2003) 827-833. https://doi.org/10.1007/s00339-002-1981-8

[30]   A. Sciuto, F. Roccaforte, S. Di Franco, V. Raineri, S. Billotta, G. Bonanno, Appl. Phys. Lett. 90 (2007) 223507. https://doi.org/10.1063/1.2745208

[31]   T.N. Oder, E. Sutphin, R. Kummari, J. Vac. Sci. Technol. B 27 (2009) 1865-1869. https://doi.org/10.1116/1.3151831

[32]   N. Kwietniewski, M. Sochacki, J. Szmidt, M. Guziewicz, E. Kaminska, A. Piotrowska, Appl. Surf. Sci. 254 (2008) 8106-8110. https://doi.org/10.1016/j.apsusc.2008.03.018

Advancing Silicon Carbide Electronics Technology I   Materials Research Forum LLC
Materials Research Foundations **37** (2018)   doi: http://dx.doi.org/10.21741/9781945291852

[33] L. Stöber, J.P. Konrath, F. Patocka, M. Schneider, U. Schmid, IEEE Transactions
      on Electron Devices 63 (2016) 578-583.
      https://doi.org/10.1109/TED.2015.2504604

[34] J.H. Zhao, K. Sheng, R.C. Lebron-Velilla, Silicon Carbide Schottky Barrier Diode,
      in "SiC Materials and Devices", M.S. Shur, M. Levinshtein, S. Rumyantsev edt.,
      pagg. 117-162, World Scientific, Singapore (2006). ISBN 981-256-835-2.
      https://doi.org/10.1142/9789812773371_0004

[35] F. Roccaforte, F. Giannazzo, V. Raineri, J. Phys. D: Appl. Phys. 43 (2010)
      223001. https://doi.org/10.1088/0022-3727/43/22/223001

[36] R. Raghunathan, B.J. Baliga, IEEE Electron Dev. Lett. 19 (1998) 71-73.
      https://doi.org/10.1109/55.661168

[37] A.L. Syrkin, J.M. Bluet, G. Bastide, T. Breatagnon, A.A. Lebedev, M.G.
      Ratsegaeva, N.S. Savkina, V.E. Chelnokov, Mater. Sci. Eng. B 46 (1997) 236-239.
      https://doi.org/10.1016/S0921-5107(96)01978-2

[38] S.K. Lee, C.M. Zetterling, M. Östling, J. Electron. Mater. 30 (2001) 242-246.
      https://doi.org/10.1007/s11664-001-0023-1

[39] A.M. Cowley, S.M. Sze, J. Appl. Phys. 36 (1965) 3212-3220.
      https://doi.org/10.1063/1.1702952

[40] M.J. Bozack, Phys. Status Solidi b 202 (1997) 549-580.
      https://doi.org/10.1002/1521-3951(199707)202:1<549::AID-PSSB549>3.0.CO;2-6

[41] T. Kimoto, Jpn. J. Appl. Phys.54 (2015) 040103.
      https://doi.org/10.7567/JJAP.54.040103

[42] S. Kurtin, T.C. McGill, C.A. Mead, Phys. Rev. Lett. 22 (1969) 1433-1436.
      https://doi.org/10.1103/PhysRevLett.22.1433

[43] S. Hara, T. Teraji, H. Okushi, K. Kajimura, Appl. Surf. Sci. 117-l 18 (1997) 394-
      399. https://doi.org/10.1016/S0169-4332(97)80113-4

[44] H. Cho, P. Leerungnawarat, D.C. Hays, S.J. Pearton, S.N.G. Chu, R.M. Strong,
      C.M. Zetterling, M. Östling, F. Ren, Appl. Phys. Lett. 76 (2000) 739-741.
      https://doi.org/10.1063/1.125879

[45] D.J. Morrison, A.J. Pidduck, V. Moore, P.J. Wilding, K.P. Hilton, M.J. Uren,
      C.M. Johnson, N.G. Wright, A.G. O'Neill, Semicond. Sci. Technol. 15 (2000)
      1107-1114. https://doi.org/10.1088/0268-1242/15/12/302

[46]   V. Khemka, T.P. Chow, R.J. Gutman, J. Electron. Mater. 27 (1998) 1128-1135.
https://doi.org/10.1007/s11664-998-0150-z

[47]   D.J. Morrison, A.J. Pidduck, V. Moore, P.J. Wilding, K.P. Hilton, M.J. Uren, C.M.
Johnson, Mater. Sci. Forum 338–342 (2000) 1199-1202.
https://doi.org/10.4028/www.scientific.net/MSF.338-342.1199

[48]   R. Pierobon, G. Meneghesso, E. Zanoni, F. Roccaforte, F. La Via, V. Raineri,
Mater. Sci. Forum 483-485 (2005) 933-936.
https://doi.org/10.4028/www.scientific.net/MSF.483-485.933

[49]   M. Treu, R. Rupp, H. Kapels, W. Bartsch, Mater. Sci. Forum 353–356 (2001) 679-
682. https://doi.org/10.4028/www.scientific.net/MSF.353-356.679

[50]   T. Hatakeyama, T. Shinohe, Mater. Sci. Forum 389–393 (2002) 1169-1172.
https://doi.org/10.4028/www.scientific.net/MSF.389-393.1169

[51]   I. Ohdomari, K. N. Tu, J. Appl. Phys. 51 (1980) 3735-3739.
https://doi.org/10.1063/1.328160

[52]   J.L. Freeouf, T.N. Jackson, S.E. Laux, J.M. Woodall, J. Vac. Sci. Technol. 21
(1982) 570-573. https://doi.org/10.1116/1.571765

[53]   Y. P. Song, R. L. Van Meirhaeghe, W.H. Lafrère, F. Cardon, Solid-State
Electronics 29 (1986) 633-638. https://doi.org/10.1016/0038-1101(86)90145-0

[54]   R.T. Tung, Appl. Phys. Lett. 58 (1991) 2821-2823.
https://doi.org/10.1063/1.104747

[55]   J. H. Werner, H. H. Güttler, J. Appl. Phys. 69 (1991) 1522-1533.
https://doi.org/10.1063/1.347243

[56]   R. T. Tung, Physical Review B 45 (1992) 13509.
https://doi.org/10.1103/PhysRevB.45.13509

[57]   R. T. Tung, Mater. Sci. Eng. R: Reports 35 (2001) 1-138.
https://doi.org/10.1016/S0927-796X(01)00037-7

[58]   D. Defives, O. Noblanc, C. Dua, C. Brylinski, M. Barthula, F. Meyer, Materi. Sci.
Eng. B 61-62 (1999) 395-401. https://doi.org/10.1016/S0921-5107(98)00541-8

[59]   B. J. Skromme, E. Luckowski, K. Moore, M. Bhatnagar, C.E. Weitzel, T. Gehoski,
D. Ganser, J. Electron. Mater. 29 (2000) 376-383. https://doi.org/10.1007/s11664-
000-0081-9

[60]   H. J. Im, Y. Ding, J.P. Pelz, W.J. Choyke, Physical Review B. 64 (2001) 075310.
        https://doi.org/10.1103/PhysRevB.64.075310

[61]   L. Calcagno, A. Ruggiero, F. Roccaforte, F. La Via, J. Appl. Phys. 98 (2005)
        023713. https://doi.org/10.1063/1.1978969

[62]   F. Roccaforte, S. Libertino, F. Giannazzo, C. Bongiorno, F. La Via , V. Raineri, J.
        Appl. Phys. 97 (2005) 123502. https://doi.org/10.1063/1.1928328

[63]   X. Ma, P. Sadagopan, T.S. Sudarshan, Physica Status Solidi (a) 203 (2006) 643-
        650. https://doi.org/10.1002/pssa.200521017

[64]   M. E. Aydın, N. Yıldırım, A. Türüt, J. Appl. Phys. 102 (2007) 043701.
        https://doi.org/10.1063/1.2769284

[65]   I. Nikitina, K. Vassilevski, A. Horsfall, N. Wright, A.G. O'Neill, S.K Ray, K.
        Zekentes, C.M. Johnson, Semicond. Sci. Technol. 24 (2009) 055006.
        https://doi.org/10.1088/0268-1242/24/5/055006

[66]   P. M. Gammon, A. Pérez-Tomás, V. A. Shah, G. J. Roberts, M. R. Jennings, J. A.
        Covington, P.A. Mawby, J. Appl. Phys. 106 (2009) 093708.
        https://doi.org/10.1063/1.3255976

[67]   K. Sarpatwari, S. E. Mohney, O.O. Awadelkarim, J. Appl. Phys. 109 (2011)
        014510. https://doi.org/10.1063/1.3530868

[68]   K.-Y. Lee, Y.-H. Huang, IEEE Transactions on Electron Devices 59 (2012) 694-
        699. https://doi.org/10.1109/TED.2011.2181391

[69]   P. M. Gammon, et al., J. Appl. Phys. 112 (2012) 114513.
        https://doi.org/10.1063/1.4768718

[70]   D. Korucu, A. Türüt, H. Efeoglu, Physica B: Condensed Matter 414 (2013) 35-41.
        https://doi.org/10.1016/j.physb.2013.01.010

[71]   P. M. Gammon, A. Pérez-Tomás, V.A. Shah, O. Vavasour, E. Donchev, J.S. Pang,
        M. Myronov, C.A. Fisher, M.R. Jennings, D. R. Leadley, P. A. Mawby,, J. Appl.
        Phys. 114 (2013) 223704. https://doi.org/10.1063/1.4842096

[72]   R. T. Tung, Appl. Phys. Rev. 1 (2014) 011304. https://doi.org/10.1063/1.4858400

[73]   G. Brezeanu, G. Pristavu, F. Draghici, M. Badila, R. Pascu, J. Appl. Phys. 122
        (2017) 084501. https://doi.org/10.1063/1.4999296

[74]   A. N. Saxena, Surface Science 13 (1969) 151-171. https://doi.org/10.1016/0039-
        6028(69)90245-3

[75]   F. Padovani, G. Sumner, J. Appl. Phys. 36 (1965) 3744-3747.
        https://doi.org/10.1063/1.1713940

[76]   L. D. Bell W. J. Kaiser, Phys. Rev. Lett. 60 (1988) 1406-1410.
        https://doi.org/10.1103/PhysRevLett.60.1406

[77]   L. D. Bell, W. J. Kaiser, Phys. Rev. Lett. 61 (1988) 2368-2371.
        https://doi.org/10.1103/PhysRevLett.61.2368

[78]   H.-J. Im, B. Kaczer, J. P. Pelz, W. J. Choyke, Appl. Phys. Lett. 72 (1998) 839-841.
        https://doi.org/10.1063/1.120910

[79]   B. Kaczer, H.J. Im, J.P. Pelz, J. Chen, W. J. Choyke, Phys. Rev. B 57 (1998)
        4027-4032. https://doi.org/10.1103/PhysRevB.57.4027

[80]   F. Giannazzo, F. Roccaforte, V. Raineri, S.F. Liotta, Europhys. Lett. 74 (2006)
        686-692. https://doi.org/10.1209/epl/i2006-10018-8

[81]   B.J. Baliga, Silicon Carbide Power Devices, World Scientific Publishing Co. Pte.
        Ltd., Singapore 2005.

[82]   D. Alok, B.J. Baliga, P.K. McLarty, IEEE Electron Device Lett. 15 (1994) 394-
        395. https://doi.org/10.1109/55.320979

[83]   D. Alok, R. Raghunathan, B.J. Baliga, IEEE Transactions on Electron Devices 43
        (1996) 1315-1317. https://doi.org/10.1109/16.506789

[84]   D. Alok, B.J. Baliga, IEEE Transactions on Electron Devices 44 (1997) 1013-
        1017. https://doi.org/10.1109/16.585559

[85]   A. Itho, T. Kimoto, H. Matsunami, IEEE Electron Device Lett. 17 (1996) 139-141.
        https://doi.org/10.1109/55.485193

[86]   R. Weiss, L. Frey, H. Ryssel, Proc. of the 14th International Conference on Ion
        Implantation Technology, Taos, New Mexico, USA, 22-27 September 2002, pp.
        139-142. https://doi.org/10.1109/IIT.2002.1257958

[87]   A. Frazzetto, F. Giannazzo, R. Lo Nigro, V. Raineri, F. Roccaforte, J. Phys. D:
        Appl. Phys. 44 (2011) 255302. https://doi.org/10.1088/0022-3727/44/25/255302

[88]   K.V. Vassilevski, N.G. Wright, A.B. Horsfall, A.G. O'Neill, M.J. Uren, K.P.
        Hilton, A.G. Masterton, A.J. Hydes, M.C. Johnson, Semicond. Sci. Technol. 20
        (2005) 271-278. https://doi.org/10.1088/0268-1242/20/3/003

[89] R. Nipoti, F. Mancarella, F. Moscatelli, R. Rizzoli, S. Zampolli, M. Ferri, Electrochemical and Solid-State Letters 13 (2010) H432-H435. https://doi.org/10.1149/1.3491337

[90] F. Roccaforte, M. Vivona, G. Greco, R. Lo Nigro, F. Giannazzo, S. Rascunà, M. Saggio, Proc. of the International Conference on Silicon Carbide and Related Materials 2017, Washington DC, US, September 17-23, 2017, Mater. Sci. Forum (2018) in press.

[91] J. Hilsenbeck, M. Treu, R. Rupp, D. Peters, R. Elpelt, Mater. Sci. Forum 615-617 (2009) 659-662. https://doi.org/10.4028/www.scientific.net/MSF.615-617.659

[92] M. Vivona, G. Greco, F. Giannazzo, R. Lo Nigro, S. Rascunà, M. Saggio, F. Roccaforte, Semicond. Sci. Technol. 29 (2014) 075018. https://doi.org/10.1088/0268-1242/29/7/075018

[93] Y. Wang, T. Li, Y. Chen, F. Cao, Y. Liu, L. Shao, IEEE Transactions on Electron Device 59 (2012) 114-120. https://doi.org/10.1109/TED.2011.2169963

[94] A. Kinoshita, T. Ohyanagi, T. Yatsuo, K. Fukuda, H. Okumura, K. Arai, Mater. Sci. Forum 645–648 (2001) 893-896.

[95] J.H. Zhao, P. Alexandrov, L. Fursin, Z.C. Feng, M. Weiner, Electron. Lett. 38 (2002) 1389. https://doi.org/10.1049/el:20020947

[96] K. Vassilevski, I.P. Nikitina, A.B. Horsfall, N.G. Wright, C.M. Johnson, Mater. Sci. Forum 645–648 (2010) 897-900. https://doi.org/10.4028/www.scientific.net/MSF.645-648.897

[97] J. Hu, L.X. Li, P. Alexandrov, X. Wang, J.H. Zhao, Mater. Sci. Forum 600–603 (2008) 947-950. https://doi.org/10.4028/www.scientific.net/MSF.600-603.947

[98] B.A. Hull, J.J. Sumakeris, M.J. O'Loughlin, Q. Zhang, J. Richmond, A.R. Powell, E.A. Imhoff, K.D. Hobart, A. Rivera-Lopez, A.R. Hefner, IEEE Transactions on Electron Devices 55 (2008) 1864-1870. https://doi.org/10.1109/TED.2008.926655

[99] T. Nakamura, T. Miyanagi, I. Kamata, T. Jikimoto, H. Tsuchida, IEEE Electron Dev. Lett. 26 (2005) 99-101. https://doi.org/10.1109/LED.2004.841473

[100] L. Zhu, T.P. Chow, K.A. Jones, C. Scozzie, A.K. Agarwal, Mater. Sci. Forum 527–529 (2006) 1159-1162. https://doi.org/10.4028/www.scientific.net/MSF.527-529.1159

[101] C. Ota, J. Nishio, T. Hatakeyama, T. Shinohe, K. Kojima, Mater. Sci. Forum 527–529 (2006) 1175-1178. https://doi.org/10.4028/www.scientific.net/MSF.527-529.1175

[102] M. Berthou, P. Godignon, J. Montserrat, J. Millan, D. Planson, J. Electron. Mater. 40 (2011) 2355-2362. https://doi.org/10.1007/s11664-011-1774-y

[103] Q-W- Song, X-Y. Tang, H. Yuan, Y-H. Wang, Y-M. Zhang, H. Guo, R-X. Jia, H-L. Lv, Y-M. Zhang, Y-M. Zhang, Chin. Phys. B 25 (2016) 047102.

[104] Q. Wahab, T. Kimoto, A. Ellison, C. Hallin, M. Tuominen, R. Yakimova, A. Henry, J. P. Bergman, E. Janzén, Appl. Phys. Lett. 72 (1998) 445-447. https://doi.org/10.1063/1.120782

[105] Q-W. Song, Y-M. Zhang, Y-M. Zhang, Q. Zhang, H-L. Lu, Chin. Phys. B 19 (2010) 087202. https://doi.org/10.1088/1674-1056/19/8/087202

[106] P. Hongbin, C. Lin, C. Zhiming, R. Jie, Journal of Semiconductors 30 (2009) 044001. https://doi.org/10.1088/1674-4926/30/4/044001

[107] R. Rupp, R. Kern, R. Gerlach, Proc. of the 25th Int. Symp. on Power Semiconductor Devices & ICs (ISPSD2013), Kanazawa, Japan, 26-30 May 2013, pagg. 51-54. https://doi.org/10.1109/ISPSD.2013.6694396

[108] M. C. Tarplee, V. P. Madangarli, Q. Zhang, T. S. Sudarshan, IEEE Transactions on Electron Devices 48 (2001) 2659-2664. https://doi.org/10.1109/16.974686

[109] G. Brezeanu, M. Badila, B. Tudor, J. Millan, P. Godignon, F. Udrea, G. Amaratunga, A. Mihaila, IEEE Transactions on Electron Devices 48 (2001) 2148-2153. https://doi.org/10.1109/16.944209

[110] F. La Via, F. Roccaforte, S. Di Franco, V. Raineri, F. Moscatelli, A. Scorzoni, G.C. Cardinali, Mater. Sci. Forum 433-436 (2003) 827-830. https://doi.org/10.4028/www.scientific.net/MSF.433-436.827

[111] M. Brezeanu et al., Proc. Int. Symposium on Power Semiconductor Devices and ICs (ISPSD 2006), Napoli, Italy, 4-8 June 2006, pp. 73-76. https://doi.org/10.1109/ISPSD.2006.1666074

[112] T. Ayalew, A. Gehring, T. Grasser, S. Selberherr, Microelectronics Reliability 44 (2004) 1473-1478. https://doi.org/10.1016/j.microrel.2004.07.042

[113] S.N. Mohammad, F.J. Kub, C.R. Eddy, J. Vac. Sci. Technol. B 29 (2011) 021021. https://doi.org/10.1116/1.3562276

[114] H. Rong, Z. Mohammadi, Y.K. Sharma, F. Li, M.R. Jennings, P.A. Mawby, Proc. of 16th European Conference on Power Electronics and Applications (EPE'14-ECCE Europe), Lappeeenranta, Finland, 26-28 August 2014. https://doi.org/10.1109/EPE.2014.6910747

[115] Y.Pan, L.Tian, H.Wu, Y.Li, F.Yang, Microelectronic Engineering 181 (2017) 10-15. https://doi.org/10.1016/j.mee.2017.05.054

[116] A. Fissel, R. Akhtariev, W. Richter, Thin Solid Films 380 (2000) 42-45. https://doi.org/10.1016/S0040-6090(00)01525-X

[117] A. Pérez-Tomás, M. R. Jennings, M. Davis, J. A. Covington, P. A. Mawby, V. Shah, T. Grasby, J. Appl. Phys. 102 (2007) 014505. https://doi.org/10.1063/1.2752148

[118] A. Pérez-Tomás, M. R. Jennings, M. Davis, V. Shah, T. Grasby, J. A. Covington, P.A. Mawby, Microelectronics Journal 38 (2007) 1233-1237. https://doi.org/10.1016/j.mejo.2007.09.019

[119] A. Pérez-Tomás, et al., Appl. Phys. Lett. 94 (2009) 103510. https://doi.org/10.1063/1.3099018

[120] M.R. Jennings, A. Pérez-Tomás, O.J. Guy, R. Hammond, S.E. Burrows, P.M. Gammon, M. Lodzinski, J.A. Covington, P.A. Mawby, Electrochemical and Solid-State Letters 11 (2008) H306-H308. https://doi.org/10.1149/1.2976158

[121] J. Liang, S. Nishida, T. Hayashi, M. Arai, N. Shigekawa, Appl. Phys. Lett. 105 (2014) 151607. https://doi.org/10.1063/1.4898674

[122] J. Liang, S. Nishida, M. Arai, N. Shigekawa, Appl. Phys. Lett. 104 (2014) 161604. https://doi.org/10.1063/1.4873113

[123] P.M. Gammon, A. Pérez-Tomás, M.R. Jennings, G.J. Roberts, M.C. Davis, V.A. Shah, S.E. Burrows, N.R Wilson, J.A. Covington, P.A. Mawby, Appl. Phys. Lett. 93 (2008) 112104. https://doi.org/10.1063/1.2987421

[124] P.M. Gammon, A. Pérez-Tomás, M.R. Jennings, V.A. Shah, S.A. Boden, M.C. Davis, SE Burrows, N.R. Wilson, G.J. Roberts, J.A. Covington, P.A. Mawby, J. Appl. Phys. 107 (2010) 124512. https://doi.org/10.1063/1.3449057

[125] R. A. Minamisawa, A. Mihaila, I. Farkas, V. S. Teodorescu, V. V. Afanas'ev, C.-W. Hsu, E. Janzén, M Rahimo, Appl. Phys. Lett. 108 (2016) 143502. https://doi.org/10.1063/1.4945332

[126] S. Sonde, F. Giannazzo, V. Raineri, R. Yakimova, J. R. Huntzinger, A. Tiberj, J. Camassel Phys. Rev. B 80 (2009) 241406. https://doi.org/10.1103/PhysRevB.80.241406

[127] S. Shivaraman, L. H. Herman, F. Rana, J. Park, M. G. Spencer, Appl. Phys. Lett. 100 (2012) 183112. https://doi.org/10.1063/1.4711769

[128] H. Tanaka, T. Hayashi, Y. Shimoida, S. Yamagami, S. Tanimoto, M. Hoshi, Proc. of the 17th International Symposium on Power Semiconductor Devices and ICs, 2005 (ISPSD2005), Santa Barbara, CA, USA, 23-26 May, 2005, pp. 287-290 (doi: 10.1109/ISPSD.2005.1488007). https://doi.org/10.1109/ISPSD.2005.1488007

[129] N.G. Wright, A.B. Horsfall, J. Phys. D: Appl. Phys. 40 (2007) 6345-6354. https://doi.org/10.1088/0022-3727/40/20/S17

[130] F. Nava, G. Bertuccio, A. Cavallini, E. Vittone, Meas. Sci. Technol. 19 (2008) 102001. https://doi.org/10.1088/0957-0233/19/10/102001

[131] F. Roccaforte, P. Fiorenza, G. Greco, R. Lo Nigro, F. Giannazzo, F. Iucolano, M. Saggio, Microelectronic Engineering 187-188 (2018) 66-77. https://doi.org/10.1016/j.mee.2017.11.021

[132] G. Brezeanu, F. Draghici, M. Badila, F. Craciunoiu, G. Pristavu, R. Pascu, F. Bernea, Mater. Sci. Forum 778-780 (2014) 1063-1066. https://doi.org/10.4028/www.scientific.net/MSF.778-780.1063

[133] R. Pascu, G. Pristavu, G. Brezeanu, F. Draghici, M. Badila, I. Rusu, F. Craciunoiu, Mater. Sci. Forum 821-823 (2015) 436-439. https://doi.org/10.4028/www.scientific.net/MSF.821-823.436

[134] S. Rao, L. Di Benedetto, G. Pangallo, A. Rubino, S. Bellone, F.G. Della Corte, IEEE Sensors Journal 16 (2016) 6537-6542. https://doi.org/10.1109/JSEN.2016.2591067

[135] S.Rao, G. Pangallo, L. Di Benedetto, A. Rubino, G.D. Licciardo, F.G. Della Corte, Procedia Engineering 168 (2016) 1003-1006. https://doi.org/10.1016/j.proeng.2016.11.326

[136] G. Pristavu, G. Brezeanu, M. Badila, F. Draghici, R. Pascu, F. Craciunoiu I. Rusu. A. Pribeanu, Mater. Sci. Forum 897 (2017) 606-609. https://doi.org/10.4028/www.scientific.net/MSF.897.606

[137] G. Brezeanu, M. Badila, F. Draghici, R.Pascu, G.Pristavu, F. Craciunoiu, I. Rusu, Proc. of the International Semiconductor Conference (CAS) Sinaia (Romania), October 12-14, 2015, pp.3-10. (doi:10.1109/SMICND.2015.7355147)

[138] M. Mazzillo, P. Shukla, R. Mallik, M. Kumar, R. Previti, G. Di Marco, A. Sciuto, R.A. Puglisi, V. Raineri, IEEE Sensors Journal 11 (2011) 377-381. https://doi.org/10.1109/JSEN.2010.2073462

[139] M. Razeghi, A Rogalski, J. Appl. Phys. 79 (1996) 7433-7473. https://doi.org/10.1063/1.362677

[140] F. Roccaforte, P. Fiorenza, G. Greco, R. Lo Nigro, F. Giannazzo, A. Patti, M. Saggio, Phys. Status Solidi (a) 211 (2014) 2063. https://doi.org/10.1002/pssa.201300558

[141] F. Yan, X. Xin, S. Alsam, Y. Zhao, D. Franz, J.H. Zhao, M. Weiner, IEEE Journ. Quantum Electronics 40 (2004) 1315-1320. https://doi.org/10.1109/JQE.2004.833196

[142] A. Sciuto, F. Roccaforte, S. Di Franco, V. Raineri, G. Bonanno, Appl. Phys. Lett. 89 (2006) 081111. https://doi.org/10.1063/1.2337861

[143] M. Mazzillo, G. Condorelli, M.E. Castagna, G. Catania, A. Sciuto, F. Roccaforte, V. Raineri, IEEE Photonics Technol. Lett. 21 (2009) 1782-1784. https://doi.org/10.1109/LPT.2009.2033713

[144] M. Mazzillo, A. Sciuto, F. Roccaforte, V. Raineri, Proc. of the 2011 IEEE Nuclear Science Symposium and Medical Imaging Conference (NSS/MIC), Valencia (Spain), 23-29 October 2011, pp. 1642-1646. https://doi.org/10.1109/NSSMIC.2011.6154652

[145] M. Mazzillo, A. Sciuto, F. Roccaforte, C. Bongiorno, R. Modica, S. Marchese, P. Badalà, D. Calì, F. Patanè, B. Carbone, A. Russo, S. Coffa, Mater. Sci. Forum 858 (2016) 1015-1018. https://doi.org/10.4028/www.scientific.net/MSF.858.1015

Advancing Silicon Carbide Electronics Technology I   Materials Research Forum LLC
Materials Research Foundations **37** (2018)   doi: http://dx.doi.org/10.21741/9781945291852

CHAPTER 4

Status and Prospects of SiC Power Devices

M. Bakowski

RISE Acreo, Isafjordsgatan 22, 164 40 Kista, Sweden

Mietek.bakowski@ri.se

Abstract

SiC power devices offer significant benefits of improved efficiency, dynamic performance and reliability in energy conversion systems. The challenges and prospects of different types of SiC devices including material and technology constraints on the device performance are reviewed. The on-state voltage and on-resistance of SiC unipolar and bipolar devices in the voltage range up to 30 kV has been determined by device simulations. System benefits and remaining challenges of SiC power electronics are summarized. Major reliability challenges of SiC power devices are reviewed and exemplified.

Keywords

Silicon Carbide, Power Devices, Unipolar, Bipolar, System Benefits, Status, Trends, Reliability

Contents

1.   Introduction

The success story of SiC power devices originated some 25 years ago and during a short period of the last 10 years a tremendous and rapid progress has been made in the field of both SiC and GaN devices. We are witnessing and participating in a technical revolution in power electronics made possible by the new WBG devices. Efficient energy conversion based on the WBG devices is necessary to master the challenges of increasing demand for electric energy and climate change. Future electric energy system requires ICT technology to control complex energy flows and ICT intensive society requires energy efficient power electronics to satisfy electric energy needs. Necessity of efficient energy conversion drives also development of new materials and technologies for future electric energy system. Efficient power electronics is thus a key technology for sustainable development.

2.   Material and technology constraints

The potential for revolutionary improvements in power device performance facilitated by availability of SiC wafers is related to some key material parameters of SiC. The most important of these are critical value of electric field (E_c) and thermal conductivity. SiC forms many different polytypes only three of which, two hexagonal 4H- and 6H-SiC and the cubic one, 3C-SiC, are of interest for power devices today. They all differ in material properties related to different bandgap values. The clearly dominant for state of the art power devices is the 4H-SiC.

The merits of SiC devices have to be constantly evaluated in relation to devices based on Si and other WBG materials being developed (GaN, Ga_2O_3, AlN, diamond). The most mature of these is GaN.

The silicon power devices are the dominant power devices today. Their main merits are still incomparable material quality, technological maturity and low material cost. However, Si power devices for medium and high voltage range applications (> 1000 V) have to be bipolar and have inherently higher switching losses compared to unipolar devices.

This speaks in favor of SiC devices since SiC unipolar devices offer low on-state resistance compared to the bipolar alternatives up to very high voltages (3.3-4.5 kV) due to the 10 times higher critical electric field value of the material. This facilitates medium and high voltage energy converters with very low losses at frequencies >100 kHz. State of the art GaN devices are also unipolar however the voltage range is still limited due to the lack of large size free standing bulk wafers necessary for vertical devices. GaN power devices today are unipolar lateral devices of HEMT type made of GaN material grown on silicon substrates most often lacking robust avalanche breakdown characteristics. This in general limits their operating voltages to below 1000 V. Transition between GaN and SiC power devices is today in the 600-900 V range.

SiC has also 3 times higher thermal conductivity compared to silicon. This facilitates higher power density and, considering 3x larger bandgap, makes SiC the preferred material for higher temperature operation. Low doped, thick epilayers with reasonably high lifetime on large area substrates (up to 8") are today feasible making SiC the preferred choice for ultra-high voltage (>10 kV) devices.

In summary; (a) SiC has 10× higher E_c compared to silicon, which means that 10× thinner drift region thickness is required for the same voltage, $V_B \approx E_c \cdot W_{sc}$, (b) this means in turn that 100× higher drift region doping is allowed, $W_{sc} \approx \sqrt{1/N_D}$, (c) assuming comparable carrier mobility this translates to 1000× lower conduction resistivity of the drift region, $\rho_{on} \approx W_{sc} \cdot (\mu_n \cdot N_D)^{-1}$, (d) unipolar SiC devices can be used in applications up to 3.3-4.5 kV resulting in extremely low conduction and switching losses.

Concerning thermal properties; (a) disregarding metallization and packaging, SiC devices could have almost 10 times higher maximum junction temperature, T_{jmax}, compared to silicon (material limit given by the thermal generation of carriers and loss of blocking properties) and (b) could handle power densities more than 10 times higher than silicon devices. The thermal limits are however set by the packaging and converter construction technologies. Operating temperatures of 200-250 °C seem to be feasible in a relatively short time span.

Epitaxy is the dominant technology in fabrication of the SiC devices. It is used for the drift layer and is the preferred technology for the main *p-n* junctions where high injection efficiency is required. This is due to the fact that annealing of implantation defects is more difficult in SiC compared to the Si. Implantation is used mostly for *p*-body in MOSFETs, *p*-type regions in JBS diodes and for junction termination.

2.1   Substrates and epi-layers

The cost of material remains the main single reason for high price of the SiC devices compared to silicon ones. The difference in €/Ampere has been shrinking with time with the increase of the wafer diameter which has increased from about 1 inch, 25 years ago, to about 8 inch today, and with devices being produced mostly on the 4 and 6 inch material.

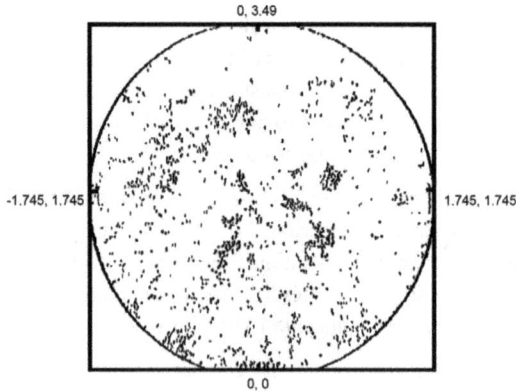

Figure 1. Typical 35mm dia. 4H-SiC wafer from 1994 with etched micropipes (courtesy of Linköping Univ.).

The device cost is though still factor 2-5 higher compared to the Si alternatives and the material cost is about 40% of the device cost for SiC MOSFETs [1]. Also, growth rates and the quality of the substrates have been greatly improved. The major material related issues which hindered device development have been related to extended defects in the material.

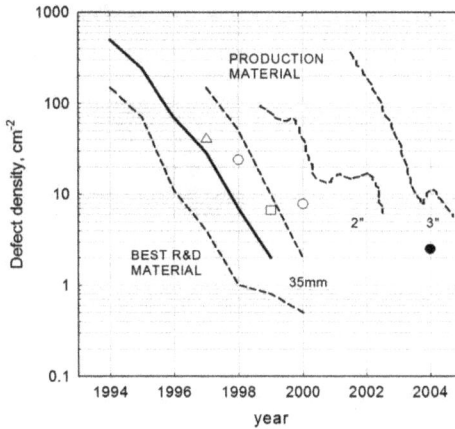

Figure 2. Defect density from yield data using Poisson model and μpipe density in Cree wafers. Published data [2] (symbols) and PiN diode data from ABB SiC project (1994-1999) (solid line).

Two historically most important have been (a) micropipes (clustered screw dislocations), a "killer" defect for blocking capability and (b) basal plane dislocations (BPDs), responsible for so called bipolar degradation which means increase in on-state voltage caused by the expansion of BPDs fueled by the charge carrier recombination energy. The impact of micropipes on the device yield is illustrated in Figs. 1 and 2 in the case of PiN and SBD rectifiers [2] and impact of BPDs on the on-state characteristics is exemplified by Fig. 3 [3,4,5,6]. The general trend visible in Fig. 2 of periodically poorer quality of wafers with each increase in diameter still applies for larger wafer diameters. The off-orientation of substrates necessary to control polytypism of the material during growth creates problems for channel mobility in MOSFETs due to the non-atomically flat surface. The difficulty of highly doped *p*-type substrates hinders development of ultra-high voltage bipolar devices like IGBT and GTO. However, many of the issues have been solved or their impact on device performance have been reduced.

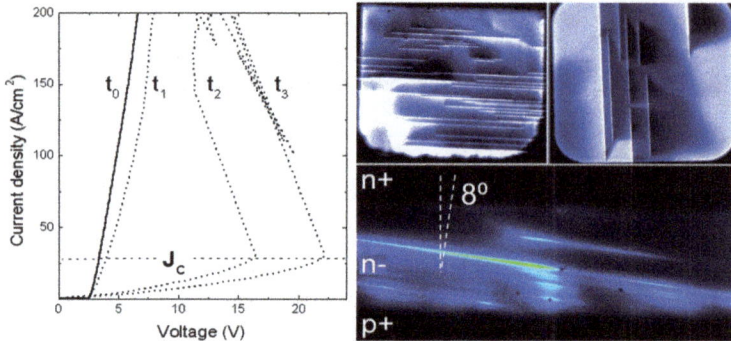

Figure 3. Increase of on-state voltage, V_{on}, during conduction due to propagation of the BPDs driven by recombination energy (left). Recombination light monitored using CCD camera (expanding bright lines) (right).

The micropipes once a major killer defect in relation to voltage blocking performance are no longer an issue even in the largest diameter crystals. Bipolar degradation is kept under control by improved quality of the material in combination with thick buffer technology used to transform basal plane dislocations into threading dislocations. The once 8 degree off orientation of the substrate material has been reduced to 4 and 2 degrees off by improvements in the growth technology. Further developments may result in further increased growth rates of the substrates and growth of thick low doped *n*-type substrates with long lifetime facilitating easy fabrication of ultra-high voltage bipolar devices in a manner familiar from silicon technology (see further section 3.3).

Table 1 summarizes the main defects remaining in the state of the art epitaxial layers [7]. Similar numbers apply also to the substrates from different suppliers [1]. In contrast to the micropipes, BPDs, and the stacking faults (SF) the influence of the remaining extended defects, threading edge dislocations (TED) and threading screw dislocations (TSD), on the device performance is relatively smaller influencing the leakage current levels and long term reliability.

Table 1. Summary of major defects in epitaxial layers [7].

Extended defects	Density [cm^{-2}]
Micropipe	0 - 0.02
BPD	0.1 - 10
TED	2000 - 5000
TSD	300 - 1000
SF	1

3.   Device types and properties

During the recent 25 years of development major device concepts known from the Si technology have been demonstrated and evaluated in silicon carbide. Fig. 4 summarizes the development and reflects the technological challenges involved as well as the impact of market preferences.

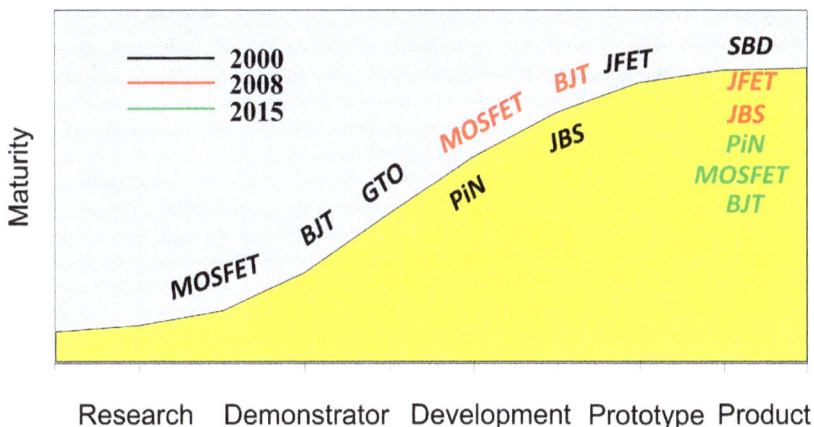

Figure 4. Schematic illustration of maturity of major device concepts over time.

Advantages and challenges of main device concepts that has been demonstrated and developed since 1990-ties are summarized in the following chapters. The gradual improvement of device performance up to now is illustrated in the technology graphs showing specific on-resistance as a function of design voltage. Instead of the specific on-resistance, ρ_{on}, the on-state voltage, V_{on}, at the specified current density is shown in many graphs in order to be able to show and compare data for both unipolar and bipolar devices in the same graph. The switching losses are not presented since they depend strongly on the driving conditions. For unipolar devices the switching losses depend on how fast we can switch the device. The switching speed is on the other hand limited by the factors external to the device itself. For motor drive applications it is the electric isolation in the windings of the electric motor and for all applications it is parasitic inductances in the packages, power modules, transformers and interconnections. Most of the diagrams are for room temperature however it should be understood that temperature is an important parameter influencing the on-state resistance and operating temperature for the application will seriously influence the maximum design voltage acceptable for a given specified maximum on-state voltage [2].

Simulation results presented in each diagram serve as a reference only and an indication of the theoretical limit of performance under idealized assumptions specified in each chapter.

3.1   Lateral channel JFET (LCJFET)

The lateral channel JFET (LCJFET) was developed and promoted by SiCED and Infineon. Schematic diagram of this device is shown in the inset in Fig. 5. The attractiveness of the concept lies in the fact that it is fabrication friendly and that device

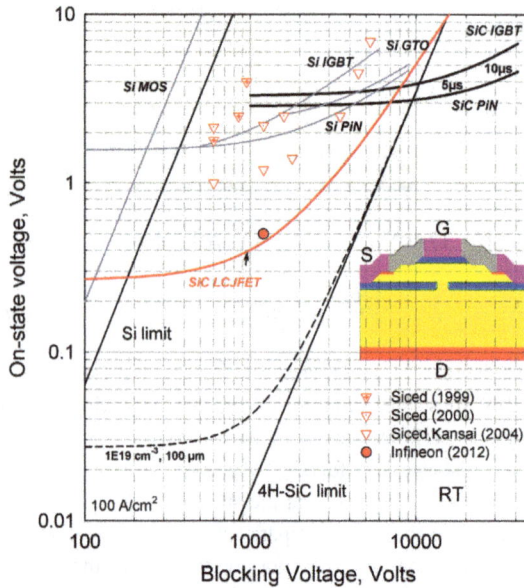

Figure 5.   Lateral channel JFET.

comprises an antiparallel body diode. The *p*-body region is made by implantation and the channel region is grown epitaxially on top. This gives a unique possibility to control the thickness and doping of the channel independently of the drift region design which provides an easy way of determining the pinch-off voltage independently of the blocking voltage.

The concept is practically purely normally-on due to the large-pitch (long channel) resulting in relatively large on-resistance and low values of saturation current. The normally-off version of the device has been developed but requires cascode configuration involving a low voltage Si MOSFET. The concept has been abandoned for switching applications with the advent of mature MOSFET devices. However, the LCJFET concept remains attractive for current limiting applications [8,9,10].

The simulated structure in Fig. 5 has cell pitch of 16 µm and threshold voltage, V_{TH}, -30 V. The data point for the latest "state of the art" device is for a structure with cell pitch of about 10 µm and V_{TH} equal to -15 V [11]. Remaining data points are from [2]. The on-state voltage is calculated for current density of 100 A/cm^2 ($\rho_{on} = V_{on}/100$ $\Omega \cdot$cm^2).

3.2 Vertical channel JFET

Different types of vertical channel JFET have been realized and evaluated. Fig. 6 shows schematically three types of JFETs: (a) Recessed Gate JFET, (b) Double Gate Buried Channel epitaxial JFET and (c) Buried Grid JFET [12].

The experimental data shown by symbols in Fig. 7 are for the devices of type (a) using the trench etching and implantation technology. The first recessed gate JFETs were produced by SemiSouth and today's commercial devices come from United Silicon Carbide (USCi).

The vertical channel JFETs of type (a) are the SiC unipolar devices with the lowest achievable ρ_{on} values. This is due to the small cell pitch and conduction controlled by the

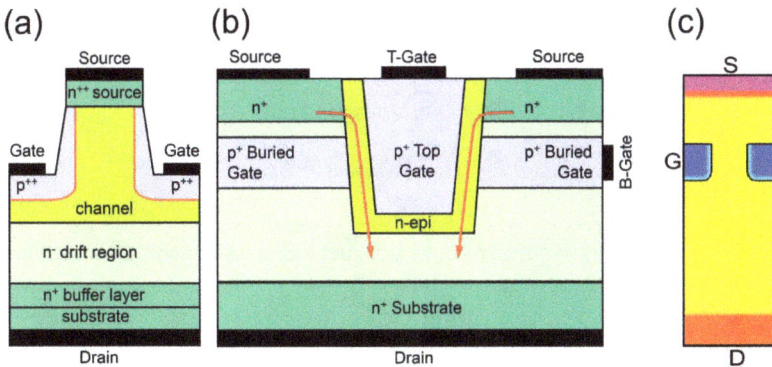

Figure 6. Schematic representations of vertical channel JFETs; Recessed Gate JFET (a), Double Gate Buried Channel JFET (b) and Buried Grid JFET (c). The structures are not to scale.

bulk mobility values. Temperature dependence of the on-resistance, R_{on}, is determined by the phonon scattering mechanism the same way as the conductivity of the bulk material and has a positive temperature coefficient which is desirable for paralleling of the devices.

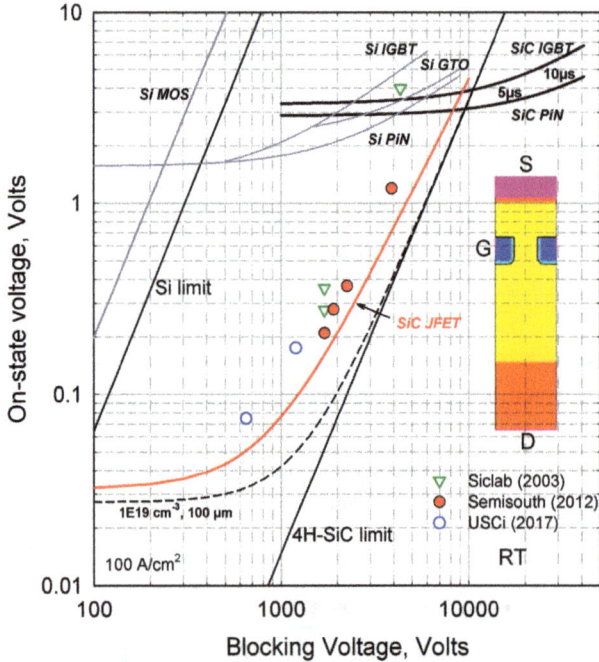

Figure 7. Vertical channel JFET.

The device is preferably Normally-on to best utilize its extremely good conduction properties. The N-off design and operation is also feasible with this concept but technological design window is very narrow and gate control requires accuracy and noise protection due to narrow range of gate control voltages typically between 1 V (V_{TH}) and 2.4 V (build-in voltage) for unipolar type of operation [12].

The N-off operation is realized by cascading the JFET with a low voltage (50 V) Si MOSFET which expands the gate control voltage range to between -25 V and +25 V alternatively 0 V and 12 V characteristic of Si MOSFET. The 650 V and 1200 V SiC cascodes are today manufactured by USCi in parallel with pure N-on vertical channel JFETs. The Si MOSFET stands for only 5 mΩ of the total cascode on-resistance [13,14]. The device has no internal body diode and normally requires external antiparallel diode in applications.

The simulation data shown by the red solid line in Fig. 7 come from a buried grid structure, shown in the inset, with a cell pitch of 5 μm, the 0.6 μm deep grid with doping of $1 \cdot 10^{19}$ cm^{-3}, grid spacing of 2 μm and top epi layer of 2 μm. The state of the art recessed gate structure from USCi has a cell pitch of 4 μm, channel length of 2.5 μm and channel width of 1.2 μm. The corresponding values for the SemiSouth structure are cell pitch 2.7 μm, channel length of 2.4 μm and channel width of 0.7 μm [15,16,17].

3.3   Bipolar SiC devices and BJT

The bipolar junction transistor is the only SiC bipolar device with very low on-state voltage even at low carrier lifetimes. The bipolar SiC devices with uneven number of *p-n* junctions suffer from high value of build-in potential due to the high bandgap of SiC. In a bipolar transistor the build-in potential of the emitter-base and collector-base junctions nearly cancel each other making the low on-state voltage possible even without conductivity modulation by injected carriers. The build-in potential is about 2.4 V for 4H-SiC at room temperature as compared to 0.6 V for silicon. This is the reason why SiC bipolar devices like PiN diode, IGBT and GTO may have lower conduction losses compared to the unipolar SiC devices first for design voltages above 4-6 kV.

The state of the art epitaxial layers still have low carrier lifetimes of the order of less than 1 μs. This gives low plasma levels in a SiC bipolar transistor resulting in a very resistive (unipolar) behavior with positive temperature coefficient of R_{on} and facilitates fast switching with low switching losses. The devices like PiN rectifier and IGBT require conductivity modulation and higher carrier lifetimes to obtain reasonable on-state voltage values. A lifetime of 1-2 μs is still acceptable for 10 kV devices, however 20 kV design voltage requires a lifetime of 5-10 μs as shown in Figs. 5,7 and 8 based on simulations. Considering the on-going progress in growth of epilayers with higher lifetimes the lifetime control becomes necessary for high frequency applications. Like for example in fast switching rectifiers for resonant power converters [18]. High temperature annealing, to control the concentration of carbon vacancies and thus uniform lifetime level, combined with local lifetime control by proton irradiation for plasma engineering may then be used, equivalent to the combination of the electron and proton irradiation in silicon power devices [19,20,21,22,23].

The main disadvantage of the BJT is that it is a current controlled device. As such it relies on the realization of as high current gain as possible. The main challenge and performance limitation of the SiC BJT concept is the surface and base contact recombination that limits the current gain. The problem is identical to what was encountered in silicon BJT devices however it is aggravated by the material related difficulties to secure good interface quality to the passivating material.

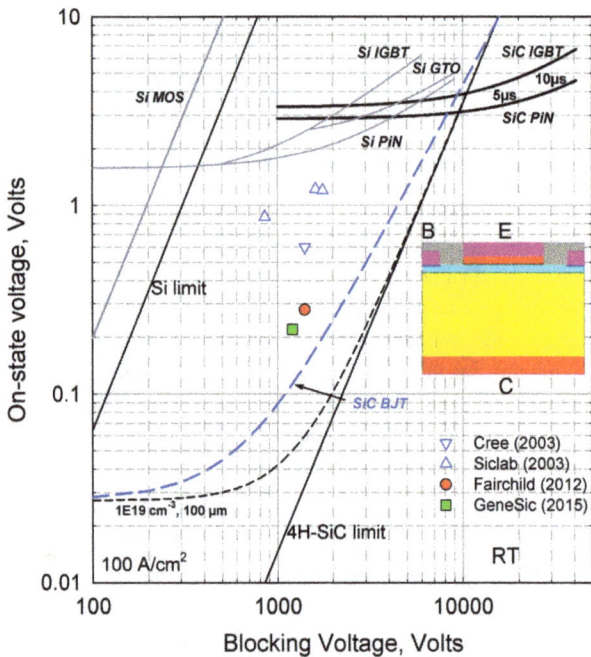

Figure 8.   *Bipolar Junction Transistor, BJT.*

The simulated structure in Fig. 8 has cell pitch of 10 μm, p-base is 1 μm thick and has a doping of $1 \cdot 10^{18}$ cm^{-3}. The lifetime in the drift region is 1 μs and the current gain, h_{FE}, of the structure is 50. The simulation data are shown by dashed blue line. The state of the art BJT structure has cell pitch of 10-20 μm, p-base thickness of 1 μm and p-base doping of $1 \cdot 10^{18}$ cm^{-3}. The lifetime in the drift region is 1 μs and the current gain, h_{FE}, of the structure is typically 50-100 [2,25,26,27].

The SJT bipolar transistors from Genesic have larger cell pitch of 66 μm (emitter width 33 μm). Normally reducing the cell pitch (emitter width) improves the switching speed of bipolar transistor but reduces the h_{FE} due to increased emitter-base junction periphery and increased loss of injected carriers due to recombination. However, in SiC BJT with low concentration of the excess carriers to be removed during switching increasing the width of the emitter and thus reducing the periphery of the base-emitter junction may offer a good choice for maximizing the current gain. The trade-off between the on-state resistance and current gain has to be considered however due to nonuniform current distribution with current density being enhanced close to the emitter edges due to the lateral *p*-base resistance (emitter current crowding effect) which speaks in favor of narrow emitter width [24]. This problem is more severe when designing SiC BJT compared to silicon due to deep energy level of acceptor dopants like aluminum resulting in so called incomplete ionization.

As a consequence of high ionization energies for acceptors (aluminum) the dopants are fully ionized in the depletion region at reverse bias but undepleted *p*-base has higher resistivity in relation to the dopant concentration at full ionization which aggravates the nonuniformity of electron injection. *P*-base dopant concentration should be high enough to be able to block design voltage and at the same time as low as possible to secure high current gain. The incomplete ionization of acceptors results in the higher lateral resistivity and attempt to compensate that by increasing the doping leads to reduction of current gain with increasing temperature. In result the trade-off between the blocking properties and current gain becomes more difficult.

The bipolar junction transistor is very robust under short circuit conditions due to current saturation in its output characteristics controlled strictly by the base current (see section 8.4).

3.4   Planar MOSFET (DMOSFET)

Inversion channel MOSFET is inherently normally-off, voltage controlled device and as such a device of choice for most applications. SiC MOSFET has been the most challenging device type to develop mainly due to the issues related to channel mobility. It took a lot of effort to bring channel mobility to acceptable values. Channel mobility values are normally in the range 3-50 cm^2/(V·sec). The main issue has been the one to two orders of magnitude higher concentration of interface states and a band of deep near interface traps (NIT) in the upper part of the bandgap [28]. In addition, SiC wafers have a nonuniform surface due to off-orientation ({0001} plane off-oriented 4 degrees towards {11$\bar{2}$0} plane) [29] necessary to prevent polytypism during crystal growth which contributes to carrier scattering during transport. Still another issue is that of the remaining implantation damage after p-body formation.

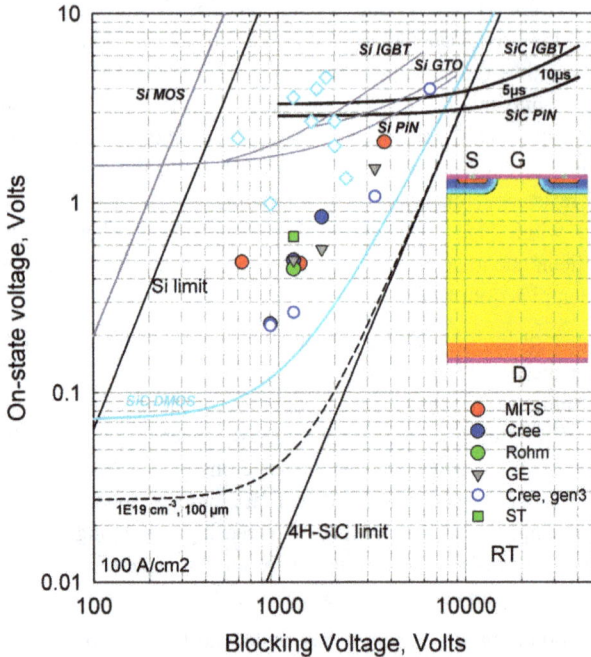

Figure 9.   The planar DMOSFET.

Due to existence of the band of deep NIT states which overlaps with the upper part of the 4H-SiC bandgap the remaining challenges of the SiC MOSFETs are low channel mobility values, low subthreshold slope, negative temperature coefficient of channel resistance and threshold voltage drift.

These are the reasons why requirements for the SiC gate driver differ from those for Si MOSFETs. In order to compensate the poor channel conduction, the SiC MOSFETs require higher positive gate voltage compared to Si MOSFETs (20-24 V as compared to standard 15 V). Also, negative gate voltage is still limited (-10 V as compared to standard -15 V) due to the threshold voltage drift under negative bias and relatively low threshold voltage values (2-3 V) for SiC MOSFETs. Low threshold voltage values are related to the necessity to improve conduction by keeping oxide thickness and p-base doping as low as possible and due to low subthreshold slope which is caused by high concentration of interface states.

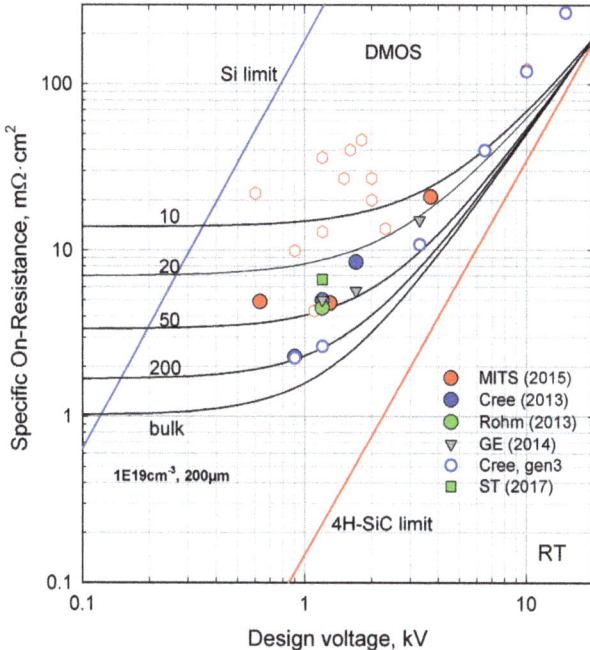

Figure 10.   Influence of channel mobility on ρ_{on} of the planar DMOSFET.

The simulated structure in Fig. 9 has cell pitch of 10 μm, channel length of 1 μm, p-body is about 1 μm deep and has a maximum doping of $1\text{-}2 \cdot 10^{18}$ cm^{-3} ($2 \cdot 10^{18}$ cm^{-3} being used for ≥ 6 kV devices). Oxide thickness is 50 nm and gate voltage is 15 V. Fig. 9 shows also the influence of the substrate thinning on the ρ_{on} assuming bulk mobility value for the channel mobility. The simulated data are shown by solid blue line. Fig. 10 shows the influence of the channel mobility on the ρ_{on} for different design voltages. Calculated lines are labeled with channel mobility values in cm^2/(V·sec).

The state of the art DMOSFET structure has cell pitch of 8 – 10 μm, channel length of about 0.5 μm, p-body is about 1 μm deep and has a maximum doping of $1 \cdot 10^{18}$ cm^{-3}. The oxide thickness is 35 – 45 nm and gate voltage is 20 V. The latest third generation

Advancing Silicon Carbide Electronics Technology I
Materials Research Foundations **37** (2018)

Materials Research Forum LLC
doi: http://dx.doi.org/10.21741/9781945291852

DMOSFET from Wolfspeed has a cell pitch of 7.5 μm resulting in increased packing density and reduced die size for the same current rating [30,31,32,33,34,35].

The performance of SiC enhancement mode MOSFETs is constantly improving due to technology development. The major breakthrough has been introduction of the nitridation and passivation of the interface traps by incorporation of atomic nitrogen into the oxide and SiC/SiO$_2$ interface. Commercial SiC MOSFETs are today available from 4 companies, however, more than 10 companies in total have SiC MOSFETs in different stages of development [1]. The leading manufacturers are Wolfspeed, ROHM and ST.

3.5 Trench MOSFET

The trench MOSFET development has as long history as planar MOSFETs, the biggest challenge has been the oxide breakdown at the edges of the trench [36]. The realization of this concept requires good control of the trench etching process and mitigation of the electric field enhancement that occurs at the trench corners. Reduction of the electric field is achieved by so called electric field shielding and can be obtained in a number of ways as discussed in section 4.2. Etching of trenches opens possibility to investigate and choose preferential crystallographic planes for the channel formation with lowest density of NIT states resulting in highest possible channel mobility [37,38,39].

The trench MOSFET concept is attractive since it facilitates small cell pitch and has no JFET resistance resulting in low values of ρ_{on} making high current density values and small chip size possible.

The advantages are the same as for silicon trench MOSFETs however main challenges are specific for SiC. Trench MOSFET shares the issues of low channel mobility caused by the high concentration of NIT states with the planar MOSFET plus the necessity to handle the RIE damage. The dielectric breakdown at the trench corners that was the main obstacle of realization of the concept is specifically related to the high critical electric field in the SiC. Since the high electric field and semiconductor breakdown normally occurs at the trench bottom corners also the field in the oxide becomes excessively high. Considering the Gauss law and the relation of dielectric constants between SiC and SiO$_2$ of 2.5 the electric field in the oxide at the trench corner obtains easily values of 6-7 MV/cm. This is too high for a reliable operation due to probability of charge injection and with respect to dielectric strength of silicon dioxide of $1 \cdot 10^7$ V/cm.

The simulated structure in Fig. 11 has cell pitch of 3 μm, channel length of 1 μm, p-body has a maximum doping of $5 \cdot 10^{17}$ cm^{-3}. Oxide thickness is 50 nm and gate voltage is 15 V. Fig. 11 shows also the influence of the substrate thinning on the ρ_{on} assuming bulk mobility value for the channel mobility. Simulated data are shown by solid green line.

Fig. 12 shows the influence of the channel mobility on the ρ_{on} for different design voltages. Calculated lines are labeled by channel mobility values in $cm^2/(V.sec)$.

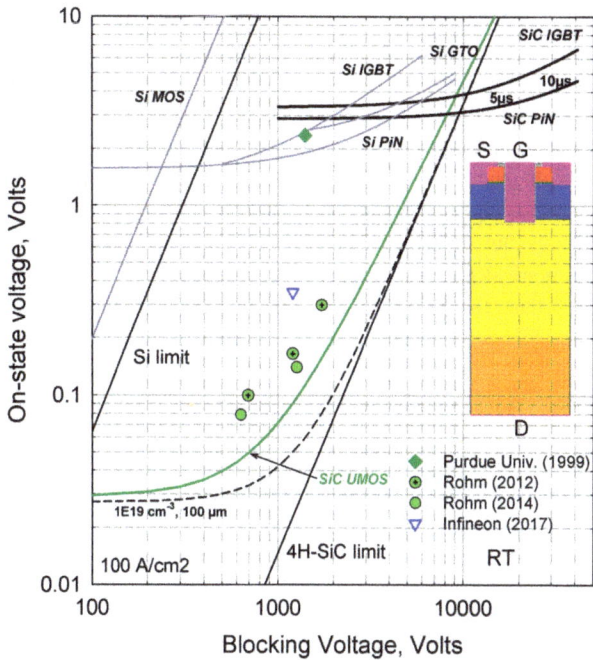

Figure 11.   The trench MOSFET.

The state of the art trench MOSFET structure has cell pitch of 4 μm, channel length of 0.5 μm, p-body has a maximum doping of $1 \cdot 10^{18}$ cm^{-3}. The oxide thickness is 40-70 nm and gate voltage is 18-20 V. Thicker oxide (>100 nm) can be used at the trench bottom at the same time for reliability reasons. Higher threshold voltage value (4-5 V) was recently realized with reasonable on-state performance at gate voltage of 15 V by Infineon [40,41,42].

4.   Performance limitations

Two selected challenges in SiC devices are illustrated by examples in the following sections. One is channel mobility in 4H-SiC and the second is electric field crowding at the trench corners which hinders full exploitation of the trench MOSFET potential for low R_{on}.

4.1 Channel mobility

Channel mobility determines the channel resistance. Channel resistivity plays more significant role in the low voltage and medium voltage MOSFETs as can be seen in Figs. 10 and 12, in sections 3.4 and 3.5. Channel mobility in SiC MOSFETs is still 5-10 times lower than bulk mobility as compared to less than factor 2 for Si MOSFETs. The reason is the high concentration of deep NIT traps and interface states which are one-two orders of magnitude higher compared to silicon [43]. As a result SiC MOSFETs have low subthreshold slope, low threshold voltages, display threshold voltage drift under gate bias and use asymmetrical gate voltages for the conduction and the off-state with positive voltage higher than for Si MOSFETs and negative voltage kept lower than for Si MOSFETs. The high positive gate voltage is to compensate for the low channel mobility. The low negative gate voltage is to prevent loss of the control since the threshold drift under negative bias is in negative direction (see section 8.3).

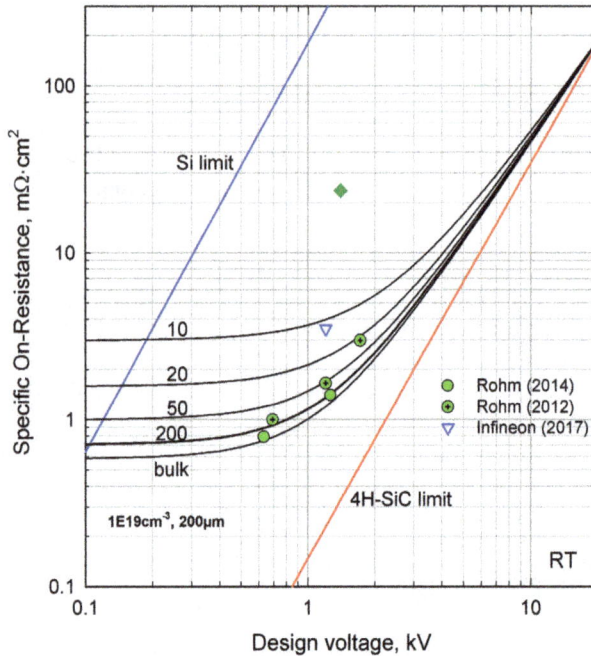

Figure 12.   Influence of channel mobility on ρ_{on} of the trench MOSFET.

Different alternative solutions have been investigated over the years to increase the channel mobility involving depletion type devices with epitaxially grown or delta doped n-channel layer, enhancement mode devices utilizing epitaxially grown p-body regions to avoid influence of implantation damage, use of other crystallographic planes for channel conduction than the normal {0001} plane, deposited dielectrics including silicon dioxide and aluminum oxide and ionic doping of SiC/SiO$_2$ interface. The largest improvement has been obtained by nitridation using atomic nitrogen, a silicon technology, which has proved efficient in passivating the interface and NIT traps.

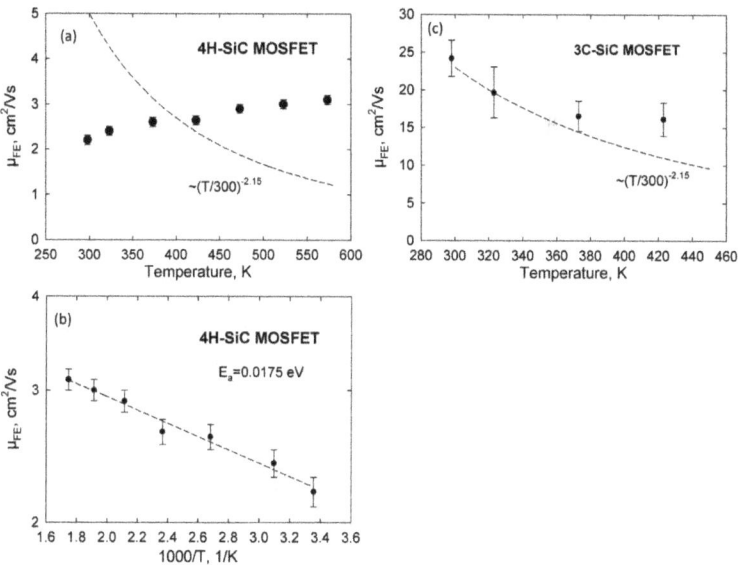

Figure 13.   Field effect channel mobility for a 4H-SiC MOSFET (left) and a 3C-SiC MOSFET (right). Note that the vertical axis in Fig. 13b has a logarithmic scale.

The influence of the NIT traps on the channel mobility and channel resistance temperature dependence in 4H-SiC is illustrated in Fig. 13. The temperature dependence of the field effect mobility is shown in Fig. 13 for a MOSFET on hexagonal 4H-SiC and cubic 3C-SiC. Close to theoretical phonon scattering dependence of mobility on temperature is observed in 3C-SiC [44,45] while the 4H-SiC shows positive temperature coefficient of mobility [46]. The hexagonal 4H-SiC polytype has a bandgap of 3.26 eV as compared to the 2.4 eV in 3C-SiC polytype. A band of deep NIT traps overlaps the upper part of the band-gap in the case of the 4H-SiC while it is situated above the bandgap in the case of the 3C-SiC [28].

This results in the hopping type of temperature activated conduction in the case of 4H-SiC as revealed by the Arrhenius plot in Fig. 13. The activation energy obtained from different devices reflects the differences between different interface treatments as well as improvements in passivation of NITs. The value of activation energy shows a clear falling trend with time and new device generations.

It is important to note that the subthreshold slope is about 5-10 times lower than for Si MOSFETs in both 3C-SiC and 4H-SiC being related to correspondingly higher density of interface states distributed over the bandgap [47]. It is a common understanding today that NIT traps are responsible for positive temperature coefficient of channel mobility, temperature activated channel conduction mechanism and threshold voltage shift (as discussed in section 8.3). While, the generally high density of both shallow and deep interface states results in low subthreshold slope [43].

The solution for enhancement mode MOSFETs is the continuous improvement of the quality of the SiC and SiO_2 interface by process and technology development.

4.2   Cell pitch in trench MOSFETs

Realization of reliable trench MOSFET requires control of the electric field enhancement at the trench corners. Different methods have been tried and developed over the years. The most straightforward is use of much thicker oxide in the trench corners and practically at the trench bottom [48]. This however has proven to be not sufficient and is often used as a complement to the electric field shielding. The electric field shielding can be achieved by implanted p-type regions either at the bottom of the trench or in the vicinity of the trench corners creating a structure of the buried grid [39,42,49]. The function of the grid is to take up the larger portion of the applied potential and thus protect the trench corner regions. The same principle has been demonstrated for buried grid JBS diodes [50,51,52]. When the reverse voltage is applied to the device, the electric field increases in all the regions including the trench corners until the n-doped regions separating the grid are fully depleted. The entire remaining voltage above that voltage value, which is determined by the structure of the grid (grid spacing) and doping of the drift region, is taken up by the grid structure alone. The structure of the p-grid can be created in the form of implanted and epitaxially overgrown p-type regions (implanted buried grid) or using the trench etching technology (recessed gate technology) and implantation into the trench walls and/or trench bottom (double trench MOSFET). The protecting trench or p-type region position must be deeper than the trench containing active MOSFET channels. The trench MOSFET from ROHM uses double trench method [42] and the most recent trench MOSFET from Infineon utilizes half of the trench for the field shielding grid and other half for the active channel [40]. All the measures utilized to control the electric field enhancement at the trench corners, involving thick oxide at trench bottom and field shielding by buried and recessed implanted p^+ regions, have a price of increased on-state resistance. This in addition to the channel mobility issues explains why the full theoretical potential of trench MOSFET with respect to the R_{on} has not been fully realized as yet. Some of the trade-offs are illustrated by Fig. 14 where a

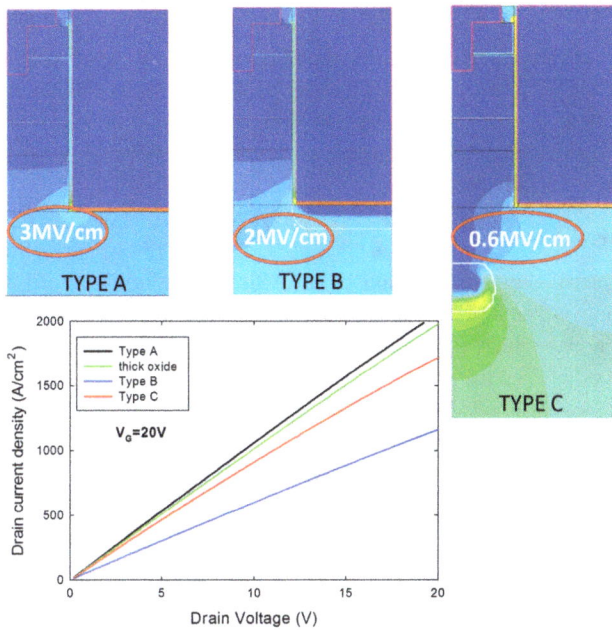

Figure 14.   Maximum electric field at the trench corner at equal applied voltage corresponding to 60% of breakdown voltage for 3.3 kV device.

comparison of two different methods of electric field shielding and their impact on the conduction characteristics is shown for the case of a 3.3 kV SiC MOSFET. Structure A has no field shielding and serves as a reference. Structure B uses p-type implantation at the bottom of the trench and structure C uses implanted buried p-type grid. A conduction characteristic for a 1 μm thick oxide at the bottom of the trench is also shown. A 33%

reduction of the electric field at the trench corner is obtained for the structure B however the on-resistance is increased by 45% compared to the reference structure A. Electric field shielding by buried grid used in structure C is more efficient resulting in reduction of electric field by 80% and smaller penalty in on-resistance which is increased by less than 20%.

5.   Material and technology curves

The on-resistance (on-state voltage) versus design voltage graphs are useful to visualize the potential of different materials and technologies for power devices. They make it also possible to compare semiconductor materials and device types. The material limit lines reflect the basic semiconductor material properties like critical value of the electric field and carrier mobility and give the resistivity of the drift region for unipolar conduction. The revolutionary concept of super-junction upsets the simple relation between the resistivity of the material and the breakdown voltage governed by the doping level and thickness of the drift region.

5.1   Super-junction concept

The principle of the super-junction is to divide the uniformly doped drift region into interchanging *n*-type and *p*-type pillars with balanced doping. The doping level in the pillars can now be made higher. During the conduction the current flows through the *n*-type doped pillars and the device resistance is reduced. At the same time, the drift region behaves as a uniform low doped region under reverse bias conditions provided the pillars are made narrow enough to be depleted. The depletion occurs now also in the lateral direction. The narrower the pillars can be made the higher doping can be used and the larger is the reduction in the on-state resistance for a given thickness of the drift region and thus given voltage. The principle of the super junction is illustrated in Fig. 15 based on numerical example of 1.2 kV design.

The super-junction concept has been realized in silicon using a combination of implantation and epitaxy, trench etching and epitaxial refill technology and trench etching and epi growth of *n*-type and *p*-type layers at trench walls (local charge balance) [53]. Different approaches have been demonstrated in SiC and GaN however the challenges and cost involved are much higher. Deep implantations are not possible in SiC due to the implantation damage, planarizations after epitaxial refill more difficult, uniformity of the doping critical for the charge balance more difficult to control. However, the Super junction concept is equally valid for SiC and has a great potential of reducing the on- resistance as shown in Fig. 16.

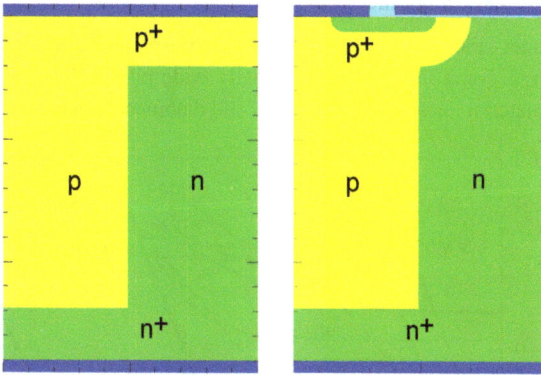

Figure 15. 1.2 kV SJ- Diode (left) and SJ-MOSFET (right). The width of the structures is 5 μm (5 μm pitch) and thickness of the drift region is 5 μm.

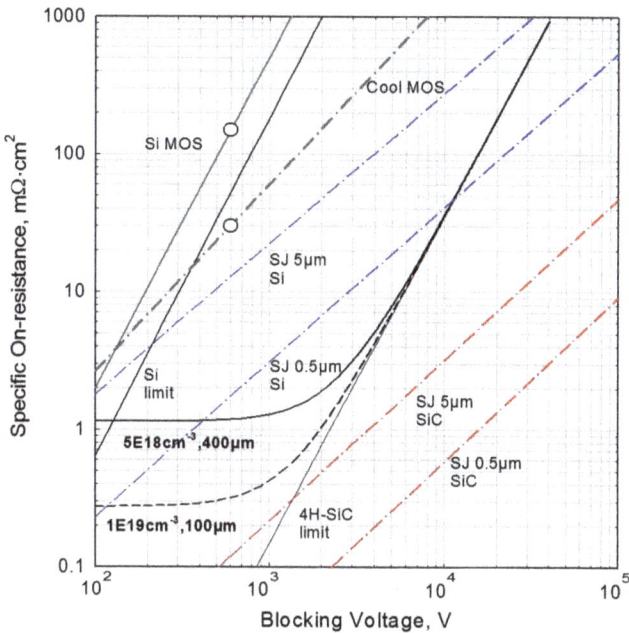

Figure 16.   Material and technology lines for SJ-design, Si and 4H-SiC.

5.2   Vertical devices using other WBG materials

The simulated super-junction technology lines are shown in Fig. 16 for both silicon and SiC for two values, 5 and 0.5 μm, of the equally wide pillars (5 and 0.5 μm pitch). A summary of the material lines for 3C-SiC, GaN and diamond is given in Fig. 17.

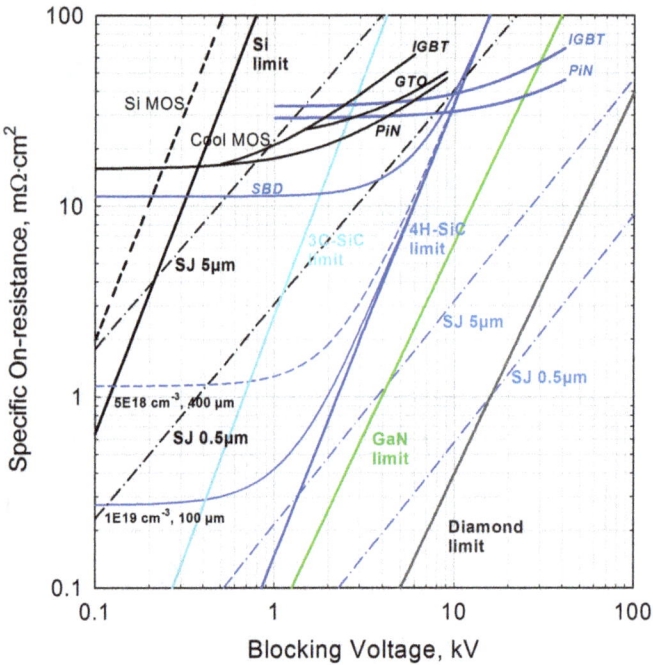

Figure 17.   Material and technology curves for SJ-design, 4H-SiC, 3C-SiC, GaN and Diamond.

The drift region of SiC devices is grown epitaxially on top of the highly doped substrate. The contribution of the substrate resistance is not negligible especially for low and medium voltage devices and is shown in addition to the pure material line calculated for the resistivity of the drift layer alone. The state of the art low and medium voltage unipolar devices use today the substrate thinning back-end technology reducing the thickness of the substrate material from the standard 300-350 μm to 100-200 μm.

6.   System benefits and applications

Efficiency of power electronic systems depends on losses in active and passive components. Efficiency of the power electronics can be greatly improved by replacing the silicon devices by silicon carbide ones. The specific material properties of SiC translate into high value added for electronic power systems. High electric breakdown field in combination with reasonably high electron mobility and high thermal conductivity translate into improved efficiency, dynamic performance and reliability of electronic and electric systems. It is relatively straight-forward to envisage savings on cooling requirements connected with increased working temperature of the devices well above the 125 to 150 °C typical of silicon power devices as well as reduced noise, size and weight of systems due to greatly increased operating frequency. To overcome both limitations has long been desirable especially in high voltage applications above 1 kV where bipolar silicon devices must be used. Such devices are necessarily slow and suffer from high switching losses due to substantial recovery charge which makes them the limiting component in the performance of many systems.

The use of SiC devices in power electronic systems opens different ways of utilization of SiC potential and new opportunities of optimization of the system. The gains of adopting SiC can mean higher power for the same system size, higher power density and reduced size, higher operating frequency and reduced size and costs of the magnetics, higher operating temperature and reduction of cooling requirements and savings on cooling sub-systems and costs. At the same time the full utilization of the SiC potential requires redesigning and adaptation of the electronic systems, development of the low inductive packaging to allow higher switching frequencies, development of packaging for higher temperatures to allow higher junction temperature, development of dedicated drivers and low inductive interconnections, control of EMC emissions and above all improvements in reliability of devices, packages and energy converters.

In short SiC devices are an obvious choice in all applications where low R_{on} (alt. low conduction losses), high frequency (viz. low switching losses) and working temperature above 150 °C (viz. reduced cooling requirements) lead to significant system benefits. The predicted benefits of introducing SiC power devices in the electronic power systems are summarized in Table 2.

Table 2. System benefits of adoption of SiC devices in power systems

Device properties	System benefits	Key applications
Low on-state voltage	Higher efficiency	Power distribution and
Low recovery charge	Higher frequency	conditioning
Fast turn-off and turn-on	Reduced noise	HVDC
High blocking voltage	Smaller size and weight	FACTS
Higher junction temperature	Smaller reactive components	Motor drives
High power density	Fewer devices in serial stack	Automotive
	Smaller heat sink or cooling by	Battery chargers
	natural convection	SMPS
	Reduced costs of forced	
	convection and fans	

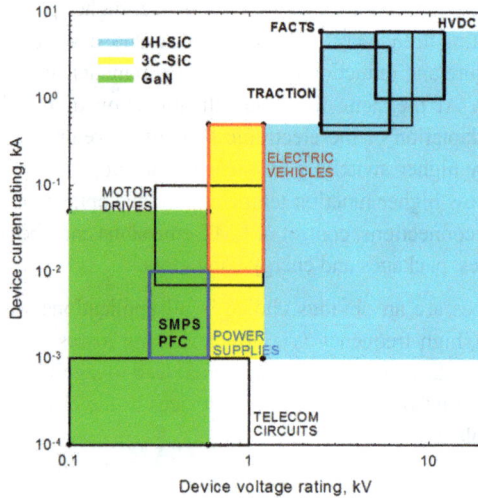

Figure 18.   Applications of WBG power devices.

The main application areas for WBG devices are schematically shown in Fig. 18. The areas where GaN and SiC devices are predicted to dominate are indicated. The intermediate voltage area indicated as suitable for 3C-SiC is an overlapping area for GaN and 4H-SiC devices. There is no commercial source of 3C-SiC wafers suitable for power device development today and the purpose is to illustrate the complementary potential of different materials and technologies as all of the above applications are still dominated by silicon devices. The final result is to a much larger extent a result of cost and availability of devices and of potential system advantages including total system costs.

7. Challenges in SiC electronics

The main driver for adoption of the SiC devices in products and applications are system advantages. In order to obtain market acceptance of the new technology it is necessary to prove reliability and reduce cost of material and devices. Cost reduction and availability of devices require access to high volume fabrication facilities on large area wafers. In order to accelerate market penetration large volume applications with relatively low demands should be targeted like SMPS (already dominant application sector) and battery chargers followed by more demanding like automotive. Wafer size and quality has to be constantly improved reducing the concentration of extended defects. Some major breakthroughs during the recent years have been nitridation improving interface quality, identification and passivation of major recombination center (carbon vacancy) by high temperature oxidation facilitating long lifetime epilayers and introduction of substrate thinning reducing the on-state resistance for medium and low voltage devices. An intensive and highly required development of low inductive and higher temperature packaging is on-going. The summary of challenges and development trends is given in Table 3.

8. Robustness and reliability

Establishing and verifying robustness and long-term reliability of WBG devices is a primary condition for adoption of new technologies in products and applications. We are at present in the initial stage of an intensive period of testing and proving reliability of SiC and GaN devices.

Of primary importance are tests related to the challenges of WBG technologies including packages. In the case of SiC power devices it is handling of high electric surface field, gate oxide reliability, threshold voltage stability, short circuit capability, power cycling, long-term blocking capability under application conditions and cosmic ray stability. All of the above are exemplified below with exception of cosmic ray reliability. The last one requires dedicated simulation effort and elaborate test environment. Also, different failure mechanisms have to be considered comparing bipolar silicon devices and unipolar SiC devices for the same application.

Table 3.   Challenges in SiC electronics

Primary	Secondary	Driver for adoption
Continuous improvement		
Wafer quality and size	Contact resistance	System benefits
Current handling capability	Resistivity of implanted layers	Device cost
MISFETs	Substrate resistivity	High volume fabrication
Channel mobility	Low leakage junction	facilities (6" and 8")
Channel engineering	termination	
Novel structures		**Reliability**
Breakthrough		
Bipolars	New material suppliers (4H-SiC)	Low threshold application
Bipolar stability	New polytypes (3C-SiC)	SMPS
Carrier lifetime (2-10 µs)	New gate dielectrics	Battery chargers
(carbon vacancy)	Substrate thickness (Substrate	Traction
Normally-off switch	thinning)	Automotive
MOSFETs		Aeronautics
Interface quality (nitridation)	Low inductive and high power	
Post-implantation damage	density module technology	Cost of energy saving
Passivation	HT packaging for T≥250 °C	
Low resistive *p*-type substrates		

8.1   Control of surface electric field

Extremely high critical electric field in SiC power devices puts severe demands on the junction termination. Generally, a termination consisting of floating field limiting rings, FLR, [54] is used for SiC devices today. This is not a space- and cost-efficient solution for medium and high voltage devices in excess of 1200 V. A large number of rings is required to obtain necessary reduction of surface electric field which makes termination to occupy a significant part of the die area. Alternative termination concept of diffused junction termination extension, JTE, [54] used widely in silicon power devices, is not feasible in SiC due to poor diffusivity of major dopants. The surface field reduction required in SiC for long term reliability is roughly one order of magnitude larger compared to silicon as long as the same passivation and isolation materials are used. An example of close to ideal junction termination for SiC devices is shown in Fig. 19.

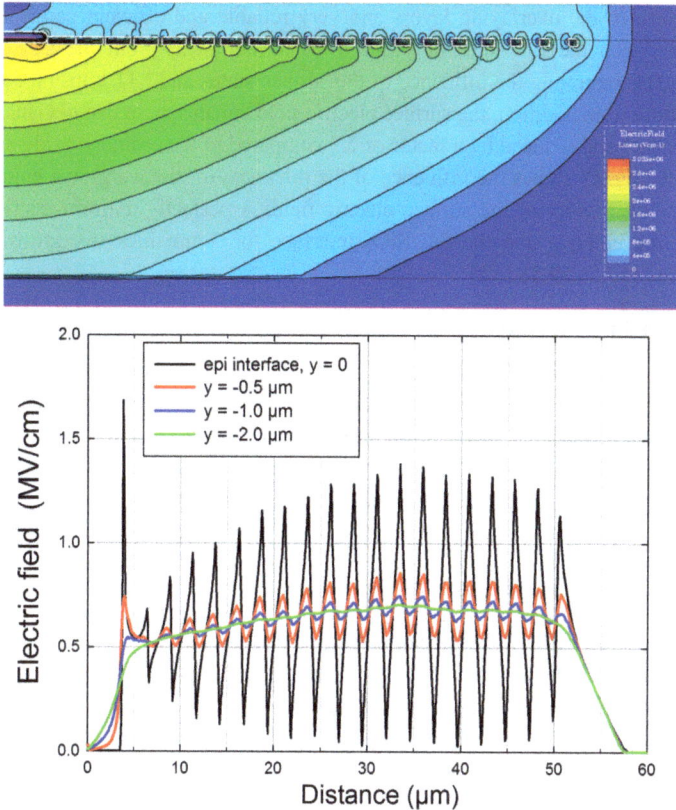

Figure 19. (a) The electric field distribution for a 3.3 kV buried junction termination with the optimum effective surface charge distribution and (b) the surface field distribution at the implanted surface (black line), 0.5 µm above (red line), 1.5 µm above (blue line) and 2 µm above the implanted surface (green line), respectively.

The termination is of the type buried JTE and is covered by a 3 µm thick low doped epilayer. The implanted regions and implantation dose are designed to give an effective

219

Advancing Silicon Carbide Electronics Technology I   Materials Research Forum LLC
Materials Research Foundations **37** (2018)   doi: http://dx.doi.org/10.21741/9781945291852

surface charge distribution according to the algorithm that yields a close to rectangular electric field distribution. This gives the most space effective termination since the blocking voltage is an integral of electric field. The implanted termination is then overgrown by epitaxial *n*-type low doped layer. This provides the termination with an ideal interface to the overlaying layers and very reliable and efficient passivation. The interface between the SiC and SiO_2 is moved further away from the active surface containing JTE removing the influence of the surface states and NIT traps on the electric field distribution. In addition, the surface electric field enhancement due to Gauss law has been eliminated and reduced by a factor of 2.5 compared to the ordinary oxide passivated surface JTE. Fig. 19 shows the influence of the thickness of the overgrown epitaxial SiC layer on the uniformity of the surface electric field. A perfectly uniform distribution at the top of the epilayer and the SiC/SiO_2 interface for 2 μm thick overgrown layer is demonstrated [55,56,57,58,59].

8.2   Gate oxide reliability

An example of a study of commercial planar and trench devices is given in Fig. 20 together with a silicon model. The devices in each group were from the same delivery. TDDB measurements were done with positive gate bias at 150 °C and 200 °C. The devices were analyzed using SEM and design parameters and oxide thickness have been extracted. The data are presented as a function of oxide field together with a lifetime model for silicon devices as described in [60]. The JEDEC TDDB model is given by:

$$t_f = A * \exp(-\gamma \cdot E_{ox}) * \exp\left(\frac{E_a}{kT}\right) \tag{1}$$

where E_{ox} is oxide field, E_a is thermal activation energy and A and γ are constants. The Fowler-Nordheim term has been neglected. The oxide thickness is 25 nm for Si devices and 40 nm and 70 nm for SiC planar and trench MOSFETs, respectively. There was a number of devices in both groups displaying extrinsic behavior with much shorter lifetimes. Only good devices are shown. The dashed lines are for E_a equal to 0.7 eV (Si) and 0.3 eV (SiC) and dash-dot lines are for E_a equal to 0.4 eV (SiC). The results indicate weaker temperature dependence in the case of SiC.

The results shown here and other published data indicate better oxide reliability for good SiC devices [35,61,62,63]. This is a very encouraging evidence considering all the interface issues. However, the issues of extrinsic failures, uniformity, reproducibility, yield and screening of devices remain forcing the users to test all the new devices and device generations.

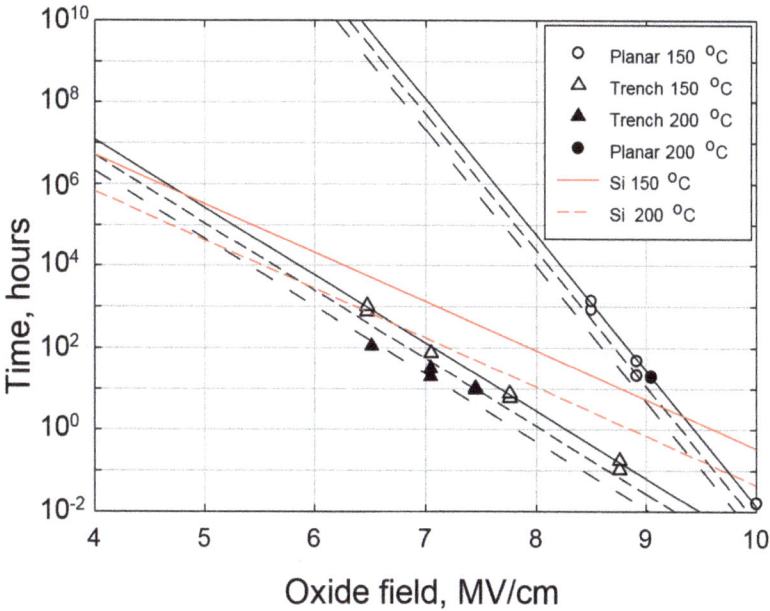

Figure 20.   TDDB data for a commercial planar DMOSFET and a trench MOSFET. JEDEC model for the Si MOSFET is shown for comparison.

8.3   Threshold voltage stability

The threshold voltage shift under negative and positive gate bias is related to charging and discharging of the NIT traps and interface states. The detailed study of the states and traps involved requires elaborate techniques to characterize time and temperature dependent phenomena. A total measure of the stability of   the commercial devices can however be obtained from the DC voltage and temperature stress with positive and negative bias for a predetermined time as shown in Fig. 21. The charge trapping at the NIT traps is by the tunneling process and results in a logarithmic time dependence as shown in Fig. 22. This allows to obtain a prediction of the total drift expected during the lifetime of the device. The choice of the current level at which V_{TH} is monitored is important since charging of NIT traps causes not only the parallel shift of the transfer characteristics but changes also the slope of the I_D vs V_G curve [64].

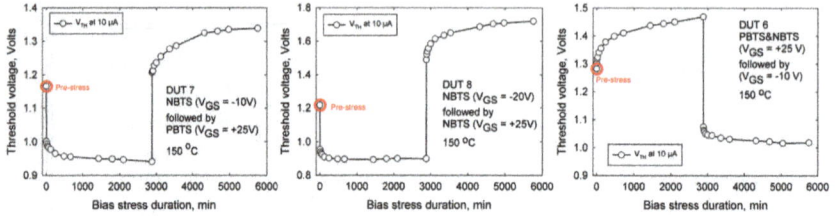

Figure 21. V_{TH} *under a DC stress with both polarities of the gate bias for 3000 min.* V_{TH} *is measured at* $I_D = 10$ *µA. Commercial DMOSFETs.*

The trapping and de-trapping of electrons at the NIT traps occurs by tunneling and the amount of V_{TH} shift depends on the concentration of traps that can be accessed by tunneling and on the concentration of electrons available for the process. Which makes the total amount of the V_{TH} shift dependent on the electric field and on the temperature. The total concentration of the NITs calculated for devices in Fig. 21 is $2\text{-}5 \cdot 10^{11}$ cm^{-2} (oxide thickness 50 nm).

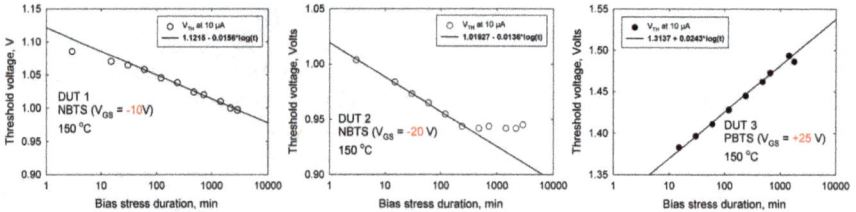

Figure 22. V_{TH} *under a DC stress with negative or positive gate bias as a function of time.* V_{TH} *is measured at* $I_D = 10$ *µA. Commercial DMOSFETs.*

8.4 Short circuit capability

Short circuit capability of power devices is an important robustness measure with respect to the application. It is important to know how long time is available to turn the device off in the case of short circuit in the system in order to prevent the destructive failure. The investigation of SiC power devices under short circuit conditions is especially important since material and electrical characteristics are distinctly different compared to the silicon devices.

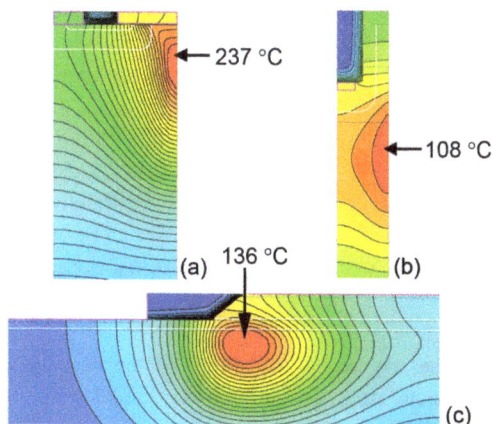

Figure 23.   Temperature distribution in the device die after a 500 ns short-circuit event (a) SiC MOSFET T_{max} = 273 °C, (b) SiC JFET T_{max} = 108 °C, (c) SiC BJT T_{max} = 136 °C. Top 10 micrometers of each structure are shown and the scale is uniform in X and Y directions. Temperature scale: 10°C per isothermal line.

The most important differences between the SiC devices can be related to the degree of saturation in device characteristics as demonstrated in Fig. 23 showing the results of short circuit behavior of commercial SiC MOSFET, JFET and BJT [65,66]. The current values corresponding to the temperature profiles in Fig. 23 measured and simulated after 0.5 µs long short circuit condition are shown in Table 4.

As can be seen in Fig. 23 the prolonged short circuit condition leads to very high temperatures close to the gate oxide and SiC interface in the SiC DMOSFET. This may contribute to the fatigue of the gate oxide and metallization and influence the long time stability of the device. Thus, a short circuit condition should be terminated as soon as possible after it is detected. The unterminated short circuit leads most probably to the melting of Al wires, arc building, discharge and destruction [66].

Table 4   Summary of peak current values after 0.5 µs of short circuit.

SiC DUT	Device type	$I_{SC,peak}$ measured [A]	$I_{SC,peak}$ simulated [A]
SCT2080KE	MOSFET	220	230
UJN1205K	JFET	120	160
GA50JT12-247	SJT (BJT)	80	100

SiC is not prone to the thermal runaway due to the wide bandgap and high temperatures required for the generation of the high enough concentration of intrinsic carriers to be comparable to the doping density in the drift or collector region. A temperature of more than 1400 °C is required to generate intrinsic carrier concentration of about $3.5 \cdot 10^{15}$ cm^{-3} [49,67] corresponding to half the doping required for the 1200 V device which is the criterion for the thermal instability [68]. SiC does not melt but sublimates which means that the unterminated short circuit will end in open circuit and explosion. This is most probable in the press-pack packages. Probability of melting of bond wires and metallization damage is much higher than reaching the sublimation temperature of about 2000 °C [69] in packages with wire bonds ending also in the open circuit failure.

8.5   Power cycling

Power cycling is a necessary test of reliability of the packaged devices testing the materials used in discrete packages and power modules as well as joining and interconnect technologies. The most probable failures are delamination, cracks in solder joints and bond wire lift-off. The risk of failures due to cracks in solder joints is increased in SiC due to 3 times higher stiffness of SiC (Young modulus) which together with 3 times higher thermal conductivity results in over 3 times higher strain energy at chip corners which could greatly accelerate solder crack propagation [70]. The most frequent and best understood failure mechanism is related to bond wire lift-off [71].

Results of power cycling of two types of commercial SiC MOSFETs (trench and planar type) are shown in Fig. 24 in comparison to Si IGBTs. All devices are packaged in TO247 packages and use wire bonding for the source electrode. The two most commonly used models, Coffin-Manson and Lesit, are plotted in the figure as well [71,72]. The Lesit model has form:

$$N_f = A * \left(\Delta T_j\right)^m * exp\left(\frac{E_a}{kT_m}\right) \tag{2}$$

where N_f is number of cycles to the fault, A is a constant, $A = 1 \cdot 10^4$, ΔT_j is temperature swing, $m = -5$, $E_a = 0.8$ eV is thermal activation energy, $T_m = \Delta T_j/2 + T_c$ is mean temperature under power cycling in Kelvin and k is Boltzmann constant. The relation between the ΔT_j and T_m with cold plate temperature $T_c = 60$ °C is given in Table 5.

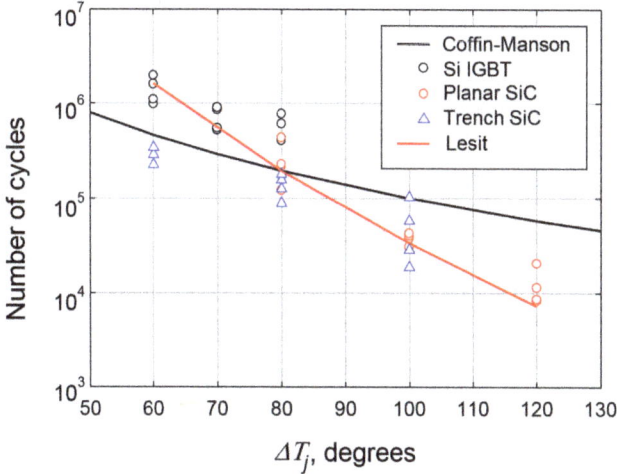

Figure 24. *Results of power cycling with different temperature swing and with cold plate temperature of 60 °C.*

Table 5. ΔT_j *and* T_m *at cold plate temperature* $T_c = 60$ °C.

ΔT_j [°C]	T_m [°C]
60	90
80	100
100	110
120	120

The tested SiC MOSFET devices have somewhat poorer lifetime compared to the reference silicon IGBTs. In addition, there are significant differences between the two MOSFET types (planar and trench) from different manufacturers. Device lifetime is however determined by the same dominant failure mechanism which is bond wire "lift-off". This has been confirmed by microscope and structural analysis after test. The results for trench MOSFET seem to be best fitted by the Coffin-Manson model and those for planar DMOSFETs are best fitted by the Lesit model. The main reason for using the Lesit model is that it is more physics of failure based since it contains a term containing thermal activation energy which could be hopefully related to the thermally activated mechanism responsible for the bond wire lift-off.

The difference between the Si and SiC devices is most probably related to the differences in contact and metallization technologies having impact on the bond strength and robustness. The differences in TCE between Si (3 ppm/ °C), SiC (3.7 ppm/ °C) and Al (24·ppm/ °C) are not significant enough to explain the difference in results.

8.6   High temperature DC storage and humidity

Results of high temperature reverse bias test with 80% reverse bias and 85% humidity at 85 °C (H3TRB) on 1200 V commercial DMOSFET devices are shown in Fig. 25. The devices show significantly increased leakage and fail after more than 1000 hrs of test. The microscopic picture of the failure areas of damaged devices is shown in Fig. 26 after decapsulation and after additional etching in H_2SO_4 to remove Al metallization.

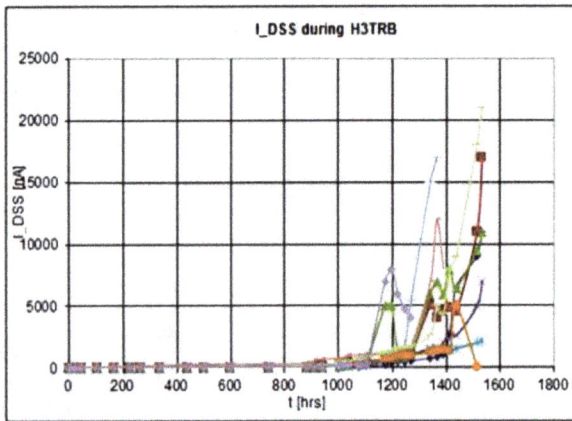

Figure 25. Results of H3TRB test at 80% of rated voltage for commercial 1200 V DMOSFET devices.

The main conclusion from the analysis is that the origin of the device failure during H3TRB test is device termination and ionic contamination. The positive ions when present in the junction termination area cause shift of the electric field distribution towards the inner part of the termination and at the same time enhancement of the electric field at the junction edge belonging to the *p*-base. The ionic contaminants, rest products of the soldering, sintering and molding processes like Sb, Sn, Na or others and oxygen ions due to dissociation of water molecules and corrosion processes are the most probable agents.

The root cause of the observed problem is a high surface electric field strength. The process is most probably accelerated in the presence of moisture due to dissociation of water molecules into H^+ and OH^- under influence of electric field. Oxygen atoms can then drift towards the outer Al ring most probably along the insulation and passivation interface causing finally delamination of the insulating film and corrosion of the Al.

The mechanism is the same as in the case of silicon devices [73]. The difference is the one order of magnitude higher surface electric field, due to the fact that passivation (SiO_2, Si_3N_4) and insulation materials (polyimide) are the same as for Si devices. The process is highly nonuniform since it is triggered by the local presence of the positive ions in the termination area. As a result, a dendritic micro-channel structure is observed locally in the passivating and insulating layers characteristic of dielectric breakdown in insulators and corrosion is clearly seen in local spots in the metallized areas. The dielectric strength of the materials is exceeded locally and as the number of the local spots is increasing with time due to the drift of the ionized species in the electric field, the blocking capability is gradually lost.

The device termination consists of FLR rings positioned in the green colored area on the left-hand side of the pictures in Fig. 26 as determined by SEM analysis. The yellow colored area to the right of the termination area is the p-base area. The first ring visible to the right of the termination area and inside the p-base area provides a short between the source and p-base. Following ring structure is part of gate metallization. The reason for failure is most probably the accumulation of positive surface charge during the long term exposure to high voltage which shifts the electric field towards the innermost FLR rings and causes local field enhancement at the edge of the p-body region closest to the shorting ring between the source and p-body. Micro-discharges are most probably part of the process and energy density dissipated in micro-discharges and conducting micro-channels may be significant even though the overall leakage current is still low. The electrical signature of the process reminds of the leakage current in the early mesa terminated SiC p-n diodes controlled by the microplasmas due to local defects [74]. Some of the microchannels become passivated and new ones are activated causing the leakage current to fluctuate as seen in Fig. 25 until large enough area of the termination becomes affected. The outcome of the process is then observed as damage to the metallization of the shorting ring. During longer exposure to high voltage the damage is extended to the gate runner and polysilicon gate and further to the closest parts of the MOSFET cells or stripes (see Fig. 26). The damage occurs in localized spots due to the nonuniform distribution of ionic species.

The long term high voltage stability is the most critical basic issue with SiC power devices. In the case of silicon power devices, the critical field strength is 0.2 MV/cm and the surface field strength is normally lowered using JTE by a factor of 3-5 with respect to the edge termination and edge passivation reliability. In the case of SiC the surface electric field strength is to start with 10 times higher while the same passivation materials are still used. That puts very severe demands on the design of the junction termination.

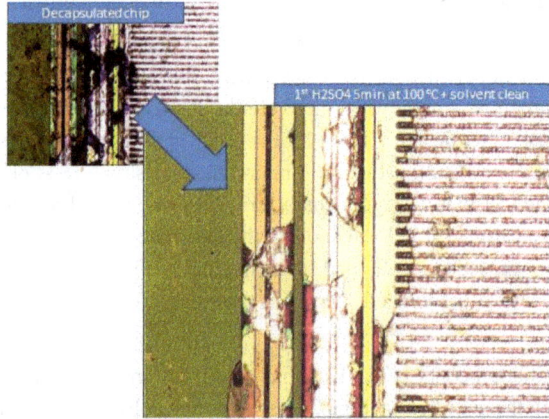

Figure 26. One of the tested devices after decapsulation and H_2SO_4 etching. One of the several damaged areas at the edge of the p-base junction. The junction termination area with floating rings is the green colored area on the left-hand side of the pictures.

Unfortunately, very straightforward solutions like field limiting rings are still used. This solution is very space demanding and can lead to sub-optimization, in an attempt to reduce the space occupied by the junction termination and thus device cost. It would be worthwhile to consider a more robust and charge tolerant design concept as demonstrated in section 8.1. Other measures are development of less water permeable molding technology and compounds and passivation materials with high dielectric constant.

9.   Conclusions and predictions

4H-SiC unipolar devices are potentially superior to all silicon devices up to 10 kV and to 4H-SiC bipolar devices up to 4-6 kV in the temperature range up to 150 °C. The low end of SiC unipolar devices is around 200-400 V provided substrate resistance is reduced by thinning the substrate down to 100 μm. The lateral GaN power devices will most probably, with time, dominate applications in the voltage range below 600-1000 V.

Remaining challenges in SiC power devices are:

(a)   bipolar instability and long carrier lifetime for high power devices above 4.5 kV and when using body diode, requiring continued improvement of material quality,

(b)   metal insulator interface quality for medium power devices in the voltage range 1.2 to 4.5 kV, requiring continued improvement to reduce the impact of channel resistance on on-state resistance, reduce the shift of threshold voltage and improve subthreshold slope,

(c)   metal insulator interface, substrate resistance, junction capacitance, to make high current density and high frequency operation possible in order to stay competitive with lateral GaN devices.

The realization of SiC potential is dependent on the market introduction of low and medium voltage SiC devices in applications like SMPS, battery chargers and automotive.

In 2-5 years perspective we can expect commercial devices with the current and voltage ratings:

(a) lateral GaN devices preferably enhancement mode $\leq$ 20 (100) A and $\leq$ 650 V (modules),

(b) unipolar SiC devices MOSFET type (modules) $\leq$ 300 (1000) A and 1200 – 3300 V,

(c) bipolar 4H-SiC devices $\geq$ 10 kV (lifetime 2-10 μs).

Acknowledgements

The author is grateful to Dr. Jang-Kwon Lim for performing the electrical measurements and structural analysis shown in sections 8.2, 8.3 and 8.6 and to Thord Nilson and Lars Lindberg, Inmotion Technologies, for the power cycling data in section 8.5.

References

[1]   Yole, Power SiC 2016 Materials Devices Modules Applications, June 2016 Report.

[2]   M. Bakowski, "Status and prospects of SiC power devices", IEEJ Transactions on Industry Applications, 126-D, (2006), no. 4, pp 391-399

[3]   A. Galeckas, Royal Institute of Technology, KTH, private communication (2000)

[4]   R. Stahlbush, N.A. Mahadik, Unexpected Sources of Basal Plane Dislocations in 4H-SiC, ECS Transactions 58 (4), 9 (2013). https://doi.org/10.1149/05804.0009ecst

[5]   N.A. Mahadik et al., Basal Plane Dislocation Mitigation Using High Temperature Annealing in 4H-SiC Epitaxy, ECS Transactions 58 (4), 325 (2013). https://doi.org/10.1149/05804.0325ecst

[6]   H. Wang et al., Studies of relaxation processes and basal plane dislocations in CVD grown homoepitaxial layers of 4H-SiC, ECS Transactions 64 (7), 213, (2014). https://doi.org/10.1149/06407.0213ecst

[7]   T. Kimoto, Material science and device physics in SiC technology for high-voltage power devices, Jpn. J. of Appl. Phys. 54 (2015) 040103. https://doi.org/10.7567/JJAP.54.040103

[8]   D. Tournier, P. Godignon, J. Montserrat, D. Planson, C. Raynaud, J. P Chante, J.-F. De Palma, F. Sarrus, A 4H-SiC high-power-density VJFET as controlled current limiter, IEEE Transactions on Industry Applications, 39(5):1508–1513, 2003. https://doi.org/10.1109/TIA.2003.816465

[9]   W. Konrad, K. Leong, K. Krischan, A. Muetze, A simple SiC JFET based AC variable current limiter, 16th European Conference on Power Electronics and Applications (EPE'14-ECCE Europe), 2014

[10]   D. Tournier, P. Godignon, S. Q. Niu, J. F. de Palma, SiC Current Limiting FETs (CLFs) for DC Applications, Materials Science Forum, Vols. 778-780, pp. 895-898, 2014. https://doi.org/10.4028/www.scientific.net/MSF.778-780.895

[11]   P. Friedrichs, Recent additions to Infineon's SiC portfolio, International SiC Power Electronics Applications Workshop ISICPEAW 2014, May 25-27, Stockholm

[12]   R.K. Malhan, M. Bakowski, Y. Takeuchi, N. Sugiyama, A. Schöner, Design, process, and performance of all-epitaxial normally-off SiC JFETs, Phys. Status Solidi A, 206, pp. 2308-2328, (2009). https://doi.org/10.1002/pssa.200925254

[13]   Streamlining Your Power Design With SiC Cascodes, APEC 2017 Seminar Sponsored by USCi, Tampa, March 26-30, 2017

[14]   C. Rockneanu, SiC Cascodes and its advantages in power electronic applications, International Wide Bandgap Power Electronics Applications Workshop, IWBGPEAW 2017, May 22-23, Stockholm

[15]   D. Sheridan, Silicon Carbide Device Update, High Megawatt Power Cond. Workshop, NIST (2012)

[16]   J.-K. Lim, D. Peftitsis, J. Rabkowski, M. Bakowski, H.-P. Nee, Analysis and experimental verification of the influence of fabrication process tolerances and circuit parasitics on transient current sharing of parallel-connected SiC JFETs, IEEE Transactions on Power Electronics, vol. 29, no. 5, pp. 2180-2191, May 2014. https://doi.org/10.1109/TPEL.2013.2281084

[17]   D-P. Sadik, J-K. Lim, J. Colmenares, M. Bakowski, H-P Nee, Comparison of Thermal Stress during Short-Circuit in Different Types of 1.2 kV SiC Transistors Based on Experiments and Simulations, Materials Science Forum, vol. 897, pp. 595-598, 2017. https://doi.org/10.4028/www.scientific.net/MSF.897.595

[18]   P. Ranstad, F. Giezendanner, M. Bakowski, J-K. Lim, G. Tolstoy, A. Ranstad, SiC Power Devices in a Soft Switching Converter including Aspects on Packaging, ECS Trans., 64 (7), p. 51 (2014). https://doi.org/10.1149/06407.0051ecst

[19]   P. Hazdra, S. Popelka, A. Schoner, Local Lifetime Control in 4H-SiC by Proton Irradiation, ICSCRM 2017, ID: 2759317

[20]   N. Thierry-Jebali, Reverse recovery control in silicon carbide high-voltage PiN diodes, International Wide BandGap Power Electronics Applications Workshop, SCAPE 2018, June 10-12, Stockholm

[21]   A. Hallén, M. Bakowski, Combined Proton and Electron Irradiation for Improved GTO Thyristors, Solid-State Electronics, Vol. 32, pp. 1033-1037, (1989). https://doi.org/10.1016/0038-1101(89)90167-6

[22]   A. Hallén, M. Bakowski, M. Lundqvist, Multiple Proton energy irradiation for improved GTO thyristors, *Solid-State* Electronics Vol. 36, No. 2, pp. 133-141, 1993. https://doi.org/10.1016/0038-1101(93)90131-9

[23]   M. Bakowski, N. Galster, A. Hallén, A. Weber, Proton Irradiation for Improved GTO Thyristors, Proc. 9th Symp. Power Semicond. Devices and ICs, Weimar (Germany), May 26-29, 1997, pp. 77-80, (1997). https://doi.org/10.1109/ISPSD.1997.601436

[24]   S. K. Ghandhi, The transistor, chapter 4.3.2, in Semiconductor Power Devices, physics of operation and fabrication technology, A Wiley-Interscience Publication, John Wiley & Sons, Inc., 1977, pp.157-162.

[25]   M. Domeij, A. Konstantinov, A. Lindgren, C. Zaring, K. Gumaelius, M. Reimark, Large area 1200 V SiC BJTs with β>100 and ρ_{ON} <3 mΩcm, Mat. Sci. Forum, vol.

717-720, pp. 1123-1126, (2012).
https://doi.org/10.4028/www.scientific.net/MSF.717-720.1123

[26] S. Sundaresan, B. Grummel, B. Hamilton, D. Singh, Improvement of the Current
Gain Stability of SiC Junction Transistors, Mater. Sci. Forum, vol. 821-823,
pp.822-825 (2015). https://doi.org/10.4028/www.scientific.net/MSF.821-823.822

[27] S. Sundaresan, B. Grummel, D. Singh, Current Gain Stability of SiC Junction
Transistors subjected to long-duration DC and Pulsed Current Stress, Mater. Sci.
Forum, vol. 858, pp.929-932 (2016).
https://doi.org/10.4028/www.scientific.net/MSF.858.929

[28] R. Schorner, P. Friedrichs, D. Peters, D. Stephani, Significantly Improved
Performance of MOSFET's on Silicon Carbide Using the 15R-SiC Polytype, IEEE
Electron Dev. Letters, vol. 20, no. 5, pp. 241- 244 (1999).
https://doi.org/10.1109/55.761027

[29] H. Matsunami, T. Kimoto, Step-controlled epitaxial growth of SiC, Mater. Aci.
Eng. R., Reports 20, pp. 125-166 (1997)

[30] J. Palmour, The era of the 2nd generation SiC MOSFET at Cree, International SiC
Power Electronics Applications Workshop ISICPEAW 2013, Stockholm, June 10-
11, 2013

[31] N. Hase, ROHM's SiC Power Device Update, Brief Introduction to MOSFET-
Only SiC Power Module and Reliability Test Results of SiC MOSFET
International SiC Power Electronics Applications Workshop ISICPEAW 2013,
Stockholm, June 10-11, 2013

[32] P. Sandvik, Progress in development and reliability of 1.2kV [& higher SiC]
devices at GE, International SiC Power Electronics Applications Workshop
ISICPEAW 2014, Stockholm, May 26-27, 2014

[33] M. Imaizumi, N. Miura, Characteristics of 600, 1200, and 3300 V Planar SiC-
MOSFETs for Energy Conversion Applications, IEEE Trans. Electron Dev., vol.
62, pp 390-395, (2015). https://doi.org/10.1109/TED.2014.2358581

[34] J. B. Casady, SiC MOSFET Commercial and Development Reliability Summary
in 2015, International SiC Power Electronics Applications Workshop ISICPEAW
2015, Stockholm, May 27-28, 2015

[35] M. Saggio, Silicon Carbide MOSFEts and Diodes for High Volume Market:
Needs, Opportunities and Perspective, International Wide BandGap Power

Electronics Applications Workshop, IWBGPEAW 2017, Stockholm, May 22-23, 2017

[36] A.K. Agarwal, R.R. Siergiej, S. Seshadri, M.H. White, P.G. McMullin, A.A. Burk, L.B. Rowland, C.D. Brandt, R.H. Hopkins, A Critical Look at the Performance Advantages and Limitations of 4H-SiC Power UMOSFET Structures, Proc. 8th Int. Symp. Power Semicond. Devices and ICs, Maui (Hawaii) May 1996, pp. 119-122.

[37] S. Onda, R. Kunmar, K. Hara, SiC Integrated MOSFETs, Physica Status Solidi (a), No. 1, pp. 369-388 (1997). https://doi.org/10.1002/1521-396X(199707)162:1<369::AID-PSSA369>3.0.CO;2-4

[38] H. Yano, H. Nakao,T. Hatayama,Y. Uraoka, T. Fuyuki, Increased Channel Mobility in 4H-SiC UMOSFETs Using On-axis Substrates, Mat Sci. Forum, Vols. 556-557, pp 807-810, (2007). https://doi.org/10.4028/www.scientific.net/MSF.556-557.807

[39] Y. Nakano, T. Mukai, R. Nakamura, T. Nakamura, A. Kamisawa, 4H-SiC Trench Metal Oxide Semiconductor Field Effect Transistors with Low On-Resistance, Japanese Journal of Applied Physics vol. 48, 04C100 (2009)

[40] Y. Nakano, R. Nakamura, H. Sakairi, S. Mitani, T. Nakamura, 690V, 1.00 mΩcm^2 4H-SiC Double-Trench MOSFETs, Mat Sci. Forum, vol. 717-720, pp. 1069-1072 (2012). https://doi.org/10.4028/www.scientific.net/MSF.717-720.1069

[41] R. Nakamura, Y. Nakano, M. Aketa, N. Kawamoto, K. Ino, 1200V SiC Trench MOSFETs, International SiC Power Electronics Applications Workshop, ISICPEAW 2014, Stockholm, May 26-27, 2014

[42] D. Peters, R. Siemieniec, T. Aichinger, T. Basler, R. Esteve, W. Bergner, D. Kueck, Performance and Ruggedness of 1200V SiC-Trench-MOSFET, Proceedings of the 29th Int. Symposium on Power Semiconductor Devices and ICs, Sapporo, (2017). https://doi.org/10.23919/ISPSD.2017.7988904

[43] V. V. Afanasev, M. Bassler, G. Pensl, M. Schulz, Intrinsic SiC/SiO$_2$ Interface States, phys. stat. sol. (a) 162, 321-337 (1997)

[44] H. Nagasawa, M. Abe, K. Yagi, T. Kawahara, N. Hatta, Fabrication of high performance 3C-SiC vertical MOSFETs by reducing planar defects, Physica Status Solidi (B) Basic Research, July 2008, 245(7), pp. 1272-1280, 2008.

[45]   M. Bakowski, A. Schöner, P. Ericsson, H. Strömberg, H. Nagasawa, M. Abe, "Development of 3C-SiC MOSFETs", Journal of telecommunications and information technology, no. 2, pp. 49-56, (2007).

[46]   S-H. Ryu et al., 950V, 8 mΩ-cm^2 High Speed 4H-SiC Power DMOSFETs, MRS 2006 Spring Meet. in San Francisco (values extracted from the presented data)

[47]   Schöner, A.; Krieger, M.; Pensl, G.; Abe, M.; Nagasawa, H. Fabrication and Characterization of 3C-SiC-Based MOSFETs, Chemical Vapor Deposition, September 2006, Vol. 12, Issue: 9, pp. 523-530, 2006.

[48]   H. Takaya, J. Morimoto, K. Hamada, T. Yamamoto, J. Sakakibara, Y. Watanabe, N. Soejima, A 4H-SiC Trench MOSFET with Thick Bottom Oxide for Improving Characteristics, Proceedings of the 25th International Symposium on Power Semiconductor Devices & ICs, Kanazawa 2013, pp. 43-46

[49]   M. Bakowski, U. Gustafsson and U. Lindefelt, Simulation of SiC High Power Devices, Phys. Stat. Sol. (a), Vol. 162, pp. 421-440, 1997. https://doi.org/10.1002/1521-396X(199707)162:1<421::AID-PSSA421>3.0.CO;2-B

[50]   M. Bakowski, HTIPM project overview, International SiC Power Electronics Applications Workshop, ISICPEAW 2007, Stockholm, March 31, 2009

[51]   M. Bakowski, J-K. Lim, W. Kaplan, A. Schöner, Merits of buried grid technology for advanced SiC device concepts, ECS Trans., 41 (8) p. 155 (2011). https://doi.org/10.1149/1.3631493

[52]   M. Bakowski, J-K. Lim, W. Kaplan, Merits of buried grid technology for SiC JBS Diodes, ECS Trans., 50 (3) p. 415 (2012). https://doi.org/10.1149/05003.0415ecst

[53]   F. Udrea, G. Deboy, T. Fujihira, Superjunction Power Devices, History, Development, and Future Prospects, IEEE Trans. on Electron Dev., vol. 64, no. 3, pp. 713-727, (2017). https://doi.org/10.1109/TED.2017.2658344

[54]   B. J. Baliga, Edge Termination, in: Power Semiconductor Devices, PWS Publishing Company, 1996, pp. 81-122.

[55]   T. Drabe, R. Sittig, Theoretical investigation of planar junction termination, Solid-State Elec., vol. 39, no. 3, pp. 323-328, Mar. 1996. https://doi.org/10.1016/0038-1101(95)00195-6

[56]   SiC semiconductor device comprising a pn junction, Mietek Bakowski, Ulf Gustafsson, Christopher I. Harris. (1999, August 3). Patent US 5,932,894 [Online]. Available: https://patentimages.storage.googleapis.com/pdfs/US5932894.pdf

[57] Fabrication of a SiC semiconductor device comprising a pn junction with a voltage absorbing edge, Mietek Bakowski, Ulf Gustafsson, Kurt Rottner, Susan Savage. (2000, March 21). *Patent US 6,040,237* [Online]. Available: https://patentimages.storage.googleapis.com/pdfs/US6040237.pdf

[58] M. Bakowski, J-K. Lim, W. Kaplan, A. Schöner,Merits of buried grid technology for advanced SiC device concepts, *ECS Trans.*, vol. 41 (8) pp. 155-158, 2011. https://doi.org/10.1149/1.3631493

[59] M. Bakowski, P. Ranstad, J-K. Lim, W. Kaplan, S. A. Reshanov, A. Schöner, F. Giezendanner, A. Ranstad, Design and Characterization of Newly Developed 10 kV 2 A SiC p-i-n Diode for Soft-Switching Industrial Power Supply, IEEE Trans on Electron Devices, Vol. 62, No. 2, p. 366, (2015). https://doi.org/10.1109/TED.2014.2361165

[60] JEDEC publication JEP122E, Failure mechanisms and models for semiconductor devices, March 2009

[61] J. Palmour, Cree – Power products reliability data and pricing forecasts for power module, power MOSFET and power diode products from 650V to 15kV, Workshop on High-Megawatt Direct-Drive Motors and Front-End Power Electronics, NIST, 2014

[62] B. Hull, D. Lichtenwalner, S-H. Ryu, E. van Brunt, J. Zhang, S. Allen, D. Grider, J. Casady, A. Burk, M. O'Loughlin, J. Palmour, Next Generation SiC MOSFETs Performance and Reliability, ARL MOS Workshop, August 18, 2016

[63] L. Stevanovic, P. Losee, S. Kennerly, A. Bolotnikov, B. Rowden, J. Smolenski, M. Harfman-Todorovic, R. Datta, S. Arthur, D. Lilienfeld, T. Schuetz, F. Carastro, F. Tao, D. Esler, R. Raju, G. Dunne, P. Cioffi, L. Yu, Readiness of SiC MOSFETs for Aerospace and Industrial Applications, Materials Science Forum, Vol. 858, pp 894-899, (2016). https://doi.org/10.4028/www.scientific.net/MSF.858.894

[64] A. J. Lelis, R. Green, D. B. Habersat and M. El, Basic Mechanisms of Threshold-Voltage Instability and Implications for Reliability Testing of SiC MOSFETs, IEEE Trans. on Electron Devices, vol. 62, no. 2, pp. 315-323, 2015. https://doi.org/10.1109/TED.2014.2356172

[65] D. P. Sadik, J-K. Lim, J. Colmenares, M. Bakowski, H-P. Nee, Comparison of Thermal Stress during Short-Circuit in Different Types of 1.2 kV SiC Transistors Based on Experiments and Simulations, Materials Science Forum, Vol. 897, pp. 897-598, 2017. https://doi.org/10.4028/www.scientific.net/MSF.897.595

[66]   D-P. Sadik, J. Colmenares, M. Bakowski, H-P. Nee, J-K Lim, Comparison of
        Thermal Stress during Short-Circuit in Different Types of 1.2 kV SiC Transistors
        Based on Experiments and Simulations, submitted to IEEE Trans on Electron
        Devices 2018.

[67]   Information on http://www.ioffe.ru/SVA/NSM/Semicond/SiC/bandstr.html

[68]   D. Silber and M. J. Robertson, Solid-State Electronics, Thermal effects on the
        forward characteristics of silicon p-i-n diodes at high pulse currents, Vol. 16, pp.
        1337-1346 (1973)

[69]   M. Syväjärvi, R. Yakimova, M. Tuominen, A. Kakanakova-Georgieva, M. F.
        MacMillan, A. Henry, Q. Wahab, E. Janzén, Growth of 6H and 4H-SiC by
        sublimation epitaxy, Journal of Crystal Growth, vol. 197, pp. 155-162, (1999).
        https://doi.org/10.1016/S0022-0248(98)00890-2

[70]   J. Lutz, R. Baburske, Some aspects on ruggedness of SiC power devices,
        Microelectronics Reliability, vol. 54, pp. 49–56, (2014).
        https://doi.org/10.1016/j.microrel.2013.09.022

[71]   J. Lutz et al., Models for lifetime Prediction, in: Semiconductor power Devices;
        Physics, Characteristics, Reliability, Springer-Verlag Berlin Heidelberg, 2011, pp.
        394-400. https://doi.org/10.1007/978-3-642-11125-9

[72]   M. Held et al., Fast Power Cycling for IGBT Modules in Traction Applications,
        Int. Conf. on Power Electronics and Drive Systems, Singapore, May 1997.
        https://doi.org/10.1109/PEDS.1997.618742

[73]   J. Lutz et al., Overvoltage - Voltage Above Blocking capability, in: Semiconductor
        power Devices; Physics, Characteristics, Reliability, Springer-Verlag Berlin
        Heidelberg, 2011, p. 431. https://doi.org/10.1007/978-3-642-11125-9

[74]   U. Zimmermann, A. Hallen, A. O. Konstantinov, B. Breitholtz, Investigation of
        Microplasma Breakdown in 4H Silicon Carbide, Materials Res. Society
        Symposium Proceedings. Vol. 512, pp. 151-156, 1998.
        https://doi.org/10.1557/PROC-512-151

Keyword Index

About the Editors

Konstantin Vasilevskiy received his MSc degree in solid-state physics from Moscow Engineering Physics Institute, USSR, in 1981. He started his research activity with characterisation of pure materials and compound semiconductors by Auger spectroscopy. In 1884, he joined Scientific-Research Institute "Orion", Kiev, USSR, where he conducted research and development on silicon IMPATT and *p-i-n* diodes. In 1988, he joined Ioffe Institute, St. Petersburg, Russia, where he started his research activity in physics and technology of wide band gap semiconductor devices. In 2002, Dr Vasilevskiy received his Ph.D. degree in physics of semiconductors for the demonstration of microwave oscillations in SiC IMPATT diodes. He is currently Senior Research Associate at the School of Engineering at Newcastle University, United Kingdom. His main research activity is focused in design, fabrication and characterization of various silicon carbide devices. He has authored over 100 scientific publications. He is a co-inventor of 16 patents granted in the field of wide band bap semiconductor technology.

Konstantinos Zekentes received his undergraduate degree in Physics, from the University of Crete, Greece, and his Ph.D., in Physics of Semiconductors, from the University of Montpellier, France. He is currently a Senior Researcher with the Microelectronics Research Group (MRG) of the Foundation for Research and Technology-Hellas (FORTH) in Heraklion, Crete, Greece and visiting researcher in the Institut de Microélectronique Electromagnétisme et Photonique et le Laboratoire d'Hyperfréquences et de Caractérisation (**IMEP-LaHC**) of CNRS/Grenoble INP/UJF/Université de Savoie. The objective of his current work is the development of SiC-related technology for elaborating high power/high frequency devices as well as SiC-based 1D devices. Dr. Zekentes has more than one hundred and seventy journal and conference publications and one US patent.

www.ingramcontent.com/pod-product-compliance
Lightning Source LLC
Chambersburg PA
CBHW071157210326
41597CB00016B/1585